AF803430

Modern Man looks at Evolution

W. W. Fletcher is Professor of Biology and past Dean of the School of Biological Sciences in the University of Strathclyde. He is a graduate of the University of Glasgow where he lectured for some time ; he has also studied at Cornell University in the United States and been a staff member of the West of Scotland Agricultural College.

He is a Fellow of the Linnean Society, Fellow and Member of Council of the Institute of Biology and Fellow of the Royal Society of Edinburgh.

He is a keen advocate of interdisciplinary studies and he makes a special study of the impact of Science and Technology on Man. He is a regular contributor to 'Science Forum' in the *Glasgow Herald* and he has written and taken part in two television series.

He is internationally known in the field of pest and pest control (in which he has published extensively in scientific papers, journals, and books). In recent years he has given a series of lectures on these topics in Holland and in Greece.

W. W. Fletcher

MODERN MAN LOOKS AT EVOLUTION

Fontana/Collins

To Marion and Guy

First published in Fontana Books 1974

© W. W. Fletcher 1974

Made and printed in Great Britain by
Love & Malcomson Ltd., Brighton Road,
Redhill, Surrey

CONDITIONS OF SALE
This book is sold subject to the condition that it shall not, by way of trade or otherwise, be lent, re-sold, hired out or otherwise circulated without the publisher's prior consent in any form of binding or cover other than that in which it is published and without a similar condition including this condition being imposed on the subsequent purchaser

Acknowledgements

This book grew out of a short course of lectures given at the American Institute of Theology at St Andrews, Scotland. One of the organizers of the Summer School, Professor William Barclay, of the Department of Divinity and Biblical Criticism, Glasgow University, persuaded me to re-write the lectures as a short book and to submit it to Collins Publishers, for consideration. Since then both Professor Barclay and Lady Collins have been constant sources of encouragement. I am very grateful to both of them. My thanks are due too to Dr Archie Roy, Senior Lecturer in Astronomy, Glasgow University, who read and criticized the first three chapters; to my colleague Professor William Hutchison who did likewise with Chapters 4–9, and to Dr Eric Perkins, Senior Lecturer in Marine Biology at Strathclyde University who made many helpful suggestions in connection with the material in Chapter 6. The final responsibility lies however with me. It is a great pleasure to acknowledge the help of my secretary, Mrs Isobel Robertson and her assistant, Mrs E. McPhail, in typing and re-typing the script, and in putting up with my many last minute corrections.

Contents

Contents

Illustrations Acknowledgements

Illustrations 1-3 are from the Larousse Encyclopedia of Astronomy and are used by kind permission of the Société encyclopédique universelle. Illustration 4 is from 'The Confirmation of Continental Drift' by Patrick M. Hurley. Copyright © 1968 by Scientific American, Inc. All rights reserved. Illustration 5 is from *Darwin and After Darwin* by G. J. Romanes and is used by kind permission of Open Court Publishing Company, Illinois. Illustration 6 is from *Biology* by Cockrum, McCauley and Younggren and is used by kind permission of W. B. Saunders Company, Philadelphia and the author. Illustrations 7 and 8 are from *Biological Sciences: Molecules to Man* (Revised Edition), Houghton Mifflin Company. By permission of the Biological Sciences Curriculum Study. The table on pages 168 and 169 is used by permission of George Weidenfeld and Nicolson Ltd.

1. The Evolution of the Universe

The size of the universe

The night is cold and frosty as we step outside the warmth of our homes and gaze into the heavens, into the great inverted bowl in which those pin-points of light which we call stars twinkle in apparently uncountable numbers. In fact no more than 5000 stars are to be seen with the naked eye on a perfectly clear night. If however we had a 4-inch lens we could see over two million and with a 200-inch mirror such as there is on Mount Palomar in the United States we could see over 1000 million.

Even modest telescopes in addition to showing millions of stars reveal the existence of diffuse patches of light. Many of these are what we call galaxies and each galaxy is composed of millions of stars.

Far out in the depths of space the innumerable atomic reactors, the stars, convert their hydrogen into helium and in the process send out streams of light that travel to us and appear as pinpoints through the vast distances of interstellar space. As we gaze we are seeing but a tiny corner of the universe, the merest tip of the galactic iceberg. Recent scientific investigations indicate that the order of size of our observable universe (and who knows how much is unobserved) is such that light, which travels at the rate of six million million miles per year, takes more than 30,000 million years to cross it. If we were to stand at the edge of our observable universe and look at a star at the other end then we would be seeing that star, not as it is now, but as it was 30,000 million years ago. We are probably seeing a 'ghost' that died and ceased to give out light millions of years ago. We may note that the light from our Sun takes eight minutes to reach us and we see our nearest star as it existed four years ago. Travelling at a rate of twelve miles per second (the speed at which our solar system speeds through space) it would take us 60,000 years to get to that nearest star.

Distances in the universe are so great that astronomers can no longer write, talk or think about them in miles. They have instead devised a new unit of measurement – the light year. One light year is the distance that light, travelling at 186,000 miles per second, travels in one year.

Galaxies

Stars are not scattered at random throughout the universe but are gathered together into groups which, as we have noted above, we call galaxies. The galaxies are often bound together in pairs e.g. our galaxy, the Milky Way, has got a twin, Andromeda, which is about two million light years away. The diameter of some of the larger galaxies is about 100,000 light years.

Clusters

Galaxies in turn occur in clusters and some such clusters may contain more than 1000 galaxies varying in size from one million to ten million light years across. The Milky Way galaxy to which our solar system belongs is a member of a cluster containing about twenty galaxies, this cluster being about three million light years in diameter.

Even the clusters do not seem randomly distributed in space. They in turn form groups of perhaps 100 clusters spread over 100 million light years. There may be more than 100,000 million galaxies in this universe and each galaxy may contain more than 100,000 million stars. The human mind accepts these figures but does not understand.

Astronomers however appear to be able to think in such

dimensions. Some of them try to help and Dr Archie Roy of Glasgow University is one of these. Take a large building, says he, such as a cathedral and call it the universe. Scatter in its interior 100,000 million specks of dust with an average separation of one third of an inch between each dust particle. Then, on a cosmic scale each dust speck represents a galaxy and each such speck contains about 100,000 million stars.

Let us magnify one of these dust specks so that we can get a better look at it – magnify it to the size of the continent of Asia and call it the Milky Way, our galaxy. Place a 10p piece on this continent and this will cover our solar system – the Sun and its attendant planets, Mercury, Venus, Earth, Mars, Jupiter, Saturn, Uranus, Neptune and Pluto. It would seem that our Earth, even our solar system, does not figure large in our universe, not even in our galaxy.

The 100,000 million stars in the Milky Way are distributed in a flattened disc which is 100,000 light years in diameter. This flattened disc rotates with the innermost point moving faster than the outer and a complete rotation takes about 200 million years. Our solar system is found near the edge of this huge disc.

Although the distances within galaxies are great, the distances between galaxies are staggering. For example the light from our neighbouring galaxy, our twin, Andromeda, which is similar in size and content to our Milky Way has been travelling through space for two million years before it reaches us.

The expanding universe

How could such a universe have come into being? The galaxies are not stationary in the universe. If we examine the light coming from any one galaxy we find that its light is reddened and its radio waves are lengthened. There is only one possible explanation for this which is generally accepted by all astronomers. It is that all the galaxies are receding from each other at enormous speeds and the further away they get then the faster they go. For example the Virgo cluster of galaxies at a distance of seven million light years from us shows a red shift indicating that it is speeding away at 600 miles per second. The most distant cluster yet identified which is 5000 million light years away is speeding off at 86,000 miles per second – 46% of the velocity of light. It is assumed that beyond this there are other galaxies so far away that they are moving close to the speed of light. By the time taken to read this sentence we will have moved a million miles away from the galaxies lying far out in the universe.

This concept of the expanding universe, of the galaxies moving apart at enormous speeds, has important implications on theories regarding the beginnings of the universe. It could be argued that if we could go back in time then the galaxies would be much closer to each other than they are now and indeed if we went back far enough they might conceivably have formed one mass of matter. Knowing the speed of recession we can then make a fair guess as to when this happened.

This calculation of recession is the basis for the 'big bang' or evolutionary theory put forward in 1927 by the Abbé Lemaître.

The evolutionary theory

Lemaître proposed that the universe originated some 20-60,000 million years ago from a dense and relatively small conglomerate of matter which was only a few million miles in diameter. This conglomerate he termed 'the primeval atom', and it contained the entire material of the universe. Its density must have been inconceivably high – about 100 million tons per cubic centimetre – and due to its instability it disintegrated in a massive cosmic explosion so that the matter making up the 'atom' was hurled out into space with such force that it continued to disperse as a vast gaseous cloud for thousands of millions of years. At last a period of stability was reached as the energy of the explosion exhausted itself and the cosmic cloud began to settle and form great clusters of galaxies. Then some 9000 million years ago instability set in again due to the forces of repulsion between the various bodies and the galaxies started to recede from each other at enormous speeds, a process which continues to this day.

The astronomer Gamow whilst accepting the idea of the primitive 'atom' (though as we shall see below having his own ideas of its composition) disagrees with Lemaître's idea that the force of the initial disintegration was exhausted after a few thousand million years. Instead he believes that the expansion of the universe which is going on today is still the result of the initial explosion and there is no need to invoke an intermediate repulsion phase. The implication of this is that the 'atom' exploded about 9000 million years ago and there is no period of diffuse gas as in the Lemaître theory. According to this 'big bang' theory, whether of Lemaître or Gamow, all the basic matter in the universe was present in the primeval 'atom' and no more is being created.

The nature of this basic matter is uncertain. Some astronomers believe that it consisted of intense radiation and corpuscular rays which formed the primeval gas during the first phase of the expansion. The energy of the primeval matter was converted first into clouds of hydrogen and then as the stars of the galaxies formed by condensation the temperature rose, converting the hydrogen into other elements. Others such as Gamow believe that the primeval atom consisted entirely of high temperature radiation.

Gamow invokes Einstein's famous equation $E=mc^2$. Before you run away from this equation let me say that it merely means that the energy contained in any matter is equal to the mass of the matter multiplied by the square of the speed of light. For our purpose it means that energy can be converted into matter and matter can be converted into energy. Here then Gamow states that at an early stage in the expansion of the 'primeval atom' (or primeval nucleus as some would prefer to call it) there was more radiant energy (light) than there was matter. But as expansion took place the density of energy thinned out faster than did the matter and at a certain point in time there was more matter than there was energy. The matter was freed from the dictatorship of energy, which formerly ensured its uniform distribution, and the gas broke up into giant clouds slowly drifting apart as the universe continued to expand. The gas clouds were cold, dark and chaotic. Gamow believes that this separation took place in the year 250,000,000 ABE (after the beginning of expansion). He calculated that the temperature of the universe was 250 million degrees centigrade when it was one hour old, dropped to 6000 degrees (present temperature of the Sun's surface) when it was 200,000 years old and had fallen to about 100 degrees below the freezing point of water when the universe reached its 250 millionth birthday. It was at this time that matter took over from energy and the galaxies began to form. On this theory all the chemical elements with which we are familiar today and whose formation depends on very high temperatures were formed within thirty minutes of the beginnings of the universe which is not so unbelievable as might appear when we realize that an atomic bomb produces a variety of materials within one microsecond of its explosion.

When however we try to visualize the pre 'primeval atom' stage we run into difficulties. Where did the 'atom' come from? As Professor Lovell the famous astronomer has said, 'It is at this stage that I feel that I have driven into a great fog barrier where the familiar world has disappeared.' Lemaître finds no great difficulty in this. For him the creation of the primeval atom was a divine act outside the limits of scientific knowledge and indeed of scientific investigation. The probable condition of intense radiation in the primeval atom is according to him entirely consistent with the divine command 'Let there be light.' Some astronomers, who believe in the theory but who are unwilling to accept a divine beginning, adopt a materialistic attitude and argue that the creation of the primeval atom has no explanation within the context of contemporary science but this does not mean that no scientific explanation will ever be possible. Others assert that the primeval atom was not the beginning but merely a state of maximum contraction of a universe which had previously existed for an eternity of time. It would appear to be logical to assume from this that contraction and expansion were continuing processes just as metal expands and contracts in heat and cold. We could call this theory that of the 'pulsating' universe. When the universe has reached a certain expansion it will begin to contract and shrink until its matter has been compressed again. It will then expand again and so on *ad infinitum*. There was no beginning and there will be no end – merely a change in the status and conditions of matter and energy which has always existed. Perhaps as humans we are conditioned to think of beginnings and ends. We are born – the beginning – and we die – the end (at least as far as the material body is concerned). This however is merely a way of looking at things. Conception and birth are a bringing together and restructuring of matter which is already in existence; death is a breakdown (in the absence of energy) of the body into its constituent chemicals. The particular structure may change – the basic materials do not.

The 'continuous creation' theory

Not all astronomers however accept the 'big bang' theory. In particular a group of British astronomers, Hoyle, Bondi, Gold and McCrea put forward an entirely new theory – that of 'continuous creation' or the 'steady-state' theory.

According to this theory the universe is the same through all space and time. However far we go back into the past, however far forward into the future, we will always find the same kind of universe as we find today. There is no beginning and no end. The universe has an infinite extent in space and no matter where one goes in the universe one would find oneself surrounded by similar galaxies. A billion years ago the universe looked as it does now, a billion years from now it will look as it does now.

In order to maintain this stable state in face of the accepted expansion of the universe these scientists postulate that the fundamental matter of the universe, hydrogen, is being continuously created at such a rate that the hydrogen atoms form galaxies to make up for those that are moving out of our field of view. This theory demands the appearance of hydrogen at the rate of several billion trillion tons per second in the observable universe but so vast is the universe that this represents the creation of only a few atoms of hydrogen per cubic mile of space per year. The authors further maintain that it is the pressure arising from the creation of this material which provides the driving force resulting in the expansion of the universe.

The question is still open as to how the universe came into being but most astronomers nowadays appear to subscribe to the 'big bang' theory. This is in large measure due to the discovery made in 1965 that the Earth is being bombarded from all directions by radiation, and indeed that the universe is filled with this radiation. It has been said that this radiation could be the remnant 'whisper' of the big bang of creation. Most former adherents of the 'steady-state' theory are now convinced that only an ancient super dense state of the universe could account for this radiation that comes in from every direction.

Further evidence against the 'steady-state' has come from the study of 'quasars'. These are starlike objects which are extremely high sources of radio energy. The combination of small size, enormous energy output, and remoteness make them objects of intense astronomical interest. It appears that they are receding at great speeds. The number of relatively faint sources of radio energy far exceed the number of strong sources. This is probably incompatible with the 'steady-state' theory.

It would appear that the 'big bang' theory is certainly the front-runner at the present time, but as in all branches of science, new observations are constantly being made and these observations, with the theories that flow from them, can sometimes topple well-established beliefs. Indeed perhaps the only certainty in science is its uncertainty. This is its strength, not its weakness as some people might suppose, since learning and knowledge advance not only by discoveries but also by scientists being willing (and sometimes forced) to give up long-cherished theories in the light of new evidence. We have an example of this in the paper written by the brilliant cosmologist Dennis Sciama on 'Cosmology before and after Quasars' (see Further Reading, *Frontiers in Astronomy*). He states, 'I must add that for me the loss of the "steady-state" theory has been a cause of great sadness. The steady-state theory has a sweep and beauty that for some unaccountable reason the architect of the Universe appears to have overlooked. The Universe in fact is a botched-job, but I suppose we will have to make the best of it.'

Formation of galaxies

It is thought that galaxies may have been formed due to disturbances and irregularities possibly due to radiations taking place in the gas cloud as it expanded, each irregularity becoming more and more dense and beginning to rotate in the form of a flat disc, thus forming a galaxy of gas. This rotating movement caused smaller disturbances to form within each galaxy, particularly towards the perimeter and these smaller disturbances each condensed to form clusters of stars. This process is a continuing one, stars are being formed even at the present day and the youngest stars may be no more than a few thousand years old. The oldest may date back 9000 million years. Some astronomers believe that not all galaxies are of the same age and that some small galaxies may even have been ejected out of larger galaxies.

As each star forms and the gaseous matter shrinks due to condensation, so high pressures are built up in its interior and as a result the temperature rises high enough to change the hydrogen, of which the gas is composed, into helium. This nuclear reaction generates great quantities of energy (as does the hydrogen bomb) which is dissipated as heat and light. As the contraction continues so the temperature continues to rise until the rate of generation of energy is the same as the rate of radiation of energy from the star's surface. A state of stability is reached, the star ceases to shrink and its size remains constant although it continues to transform hydrogen to helium and thus release energy. Our Sun, which is a star, is in this condition. Among the many stars in our galaxy are a few whose fires have gone out. They have exhausted their fuel and have shrunk to a fraction of their former size. These stars are called 'white dwarfs'. Eventually 'white dwarfs' will become 'black dwarfs' when their light no longer reaches us.

There are other types of stars such as the super-giants which are about ten times the size of the Sun. These have a relatively short existence due to the rapid turnover of their hydrogen into helium. There is an excessive loss of energy and as a result they gradually shrink and, rotating with ever increasing speed, they throw off matter into space. Such a star is termed an ordinary nova.

In some cases however the super-giants shrink very rapidly and as a result of the tremendous increase in pressure the temperature rises to such a point that the helium derived from the hydrogen is in turn converted into higher elements. At 100 to 200 million degrees centigrade the helium is converted into carbon, oxygen and neon. The temperature is still rising and at 1500 million degrees silicon, aluminium and calcium make their appearance. At 2000 million degrees and above iron, nickel, chromium and other metals are being formed in the heart of the nuclear reactor, in the core of the star. Still the temperature is rising and as a result of collapse there is a massive disintegration which throws vast clouds of incandescent matter, interspersed with the various elements, out into space. These exploding stars are termed supernovae and they may form some of the stuff of which our earth and other planets are composed. On 4 July 1054 one of these supernovae explosions took place. A star in our galaxy which had been invisible suddenly became one of the brightest in the sky – so bright that it could be seen in daylight. Then it exploded with a display of fireworks, the brightness of which Man had not seen before and has not seen since. These fireworks lasted in the day skies for three weeks and then gradually faded and in two years had disappeared from the naked eye. This happening is based on observations made in China and Japan and the Chinese named it the 'guest star'. It must have been clearly seen in Europe too but it does not appear to have been recorded.

There is a follow-up to this story because in the eighteenth century, by which time telescopes had come into general use, astronomers spotted a nebula at the location of the vanished star. This came to be known as the Crab Nebula which is still readily visible and is the debris of the explosion in 1054.

Galileo Galilei

At the centre of our planetary system is the Sun; the Earth and the planets revolve round it. This is readily accepted today but it was not always so readily accepted in the past. Aristotle taught that the stars and the planets moved around the Earth which was the centre of the universe. His views held sway for nearly 2000 years despite the fact that Aristarchus of Samos, an astronomer of the third century BC questioned them and maintained that the Earth and the planets moved round the Sun. One wonders at the intellectual capacity and indeed bravery of this man Aristarchus. He was flying in the face of the obvious. How could one believe that this massive Earth of ours was actually moving – it was contrary to all reason and direct observation. Yet Aristarchus was right, though his views lay buried until Copernicus uncovered and restated them some years later. Luther said of Copernicus 'The fool will turn the whole science of astronomy upside down.' And he did in his *De Revolutionibus*, but his views did not become widespread until some 40 years later in 1584 when Giordano Bruno published three works on the Copernican theories. For his bravery he was burned at the stake in 1600. The hunt was up however and in 1609 Galileo turned his telescope towards the heavens.

Galileo Galilei is probably the greatest of all the Italian scientists and one of the greatest of all time. He has been described as 'the father of modern science' because he took science out of the dark fog of mysticism of the Middle Ages into the logical light of observation and deduction. He set science on the pathway that it has followed ever since. He was born in Pisa in 1564 and studied medicine at the local university. He found however that there was not enough challenge in medicine for his fertile brain and he turned to experimental philosophy. After inventing the pendulum, which he used to measure his pulse, he accepted the chair in mathematics in Padua in 1592 because it is said that his revolutionary discoveries and biting satire had made him many enemies in Pisa. He stayed in Padua till 1610; among his inventions during these years were a thermometer and a compass.

He was the first to use his telescope for astronomical observations and he discovered a number of startling and brilliant facts. He found that the Milky Way was 'powdered with stars', that the moon, far from being a smooth and self-luminous sphere, had mountains and valleys and reflected the Sun's rays, that the planet Jupiter had four satellites (or moons), that Saturn due to its rings looked like three planets, and that the Sun which had always been regarded as perfect was really marked with spots.

These discoveries were sensational, although some thought that the images were ghosts. One philosopher of the time said, 'The satellites of Jupiter are invisible to the naked eye and therefore can have no influence on the Earth, and therefore would be useless, and therefore cannot exist.' Galileo's observations convinced him of the rightness of the theories of Copernicus that the planets, including the Earth, went round the Sun.

Such ideas of course brought him into sharp conflict with the Church in which the belief was firmly implanted that the Sun with her attendant planets circled the Earth. He was summoned to Rome where he was obliged to abjure his 'heresies', above all that of the daily and yearly motion of the Earth and the stability of the Sun. He was required also to recite the seven Penitential Psalms once a week for three years and he was placed under observation in a friend's palace for the rest of his life. Yet much of his best work was done during this period; he studied the nature of motion and realized that force is not necessary to keep things in motion. Many before Galileo, whilst accepting the views of Copernicus, had thought that the planets were driven round the Sun by something like giant spokes sticking out of the Sun.

Optical astronomy

For more than 300 years after Galileo first looked at the sky with his telescope progress in observational astronomy was largely concerned with the constructing of bigger and better telescopes. This culminated in the 200-inch telescope on Mount Palomar in the United States which began operating in the early 1950s and can photograph star systems whose light has taken more than 2000 million years to reach us. It appears that this may be the ultimate in optical telescopes on Earth because the unsteadiness of our atmosphere, even under the excellent conditions of Mount Palomar limits the full use of this telescope to only a few dozen nights in the year.

It is known also that the atmosphere blocks out most of the infra-red and ultra-violet rays that come from the Sun and other stars. An orbiting astronomical laboratory was launched in December 1968 and it provides information on the ultra-violet output of about 50,000 stars. In particular the Sun and the peculiar stars known as pulsars (first discovered in 1968) which give out either light or radio signals at regular intervals are being observed.

Viewed by infra-red light, relatively cool objects glow with a great brilliance. Using this technique it may be possible to observe solar systems in the making.

Radio astronomy

Another very powerful weapon which astronomers have at their command is the radio telescope. Radio waves travel at the same speed as light and in 1930 Jansky, an American engineer, found that they were reaching the Earth from outer space. Since then radio astronomy has proved to be one of the most fruitful fields of astronomical investigation. Many of these radio waves come from sources far beyond the reach of our optical telescopes. The most spectacular source is in the Crab Nebula which is 4000 light years away. This as we have noted is the gaseous remains of a supernova explosion seen by Chinese astronomers in 1054. There are indeed hundreds of radio sources, some inside and some far outside our own galaxy.

X-ray astronomy

Until very recently astronomers have had at their disposal only those wavelengths that could reach earth through the atmosphere, namely optical and radio waves. However with the development of rockets it is now possible to place machines on board that will detect x-rays and some stars in the heavens have been shown to be very powerful sources. The technique is being used to investigate many astronomical problems.

Space observatories

By 1980 astronomers hope to set up a great observatory in space. In this observatory there will be reflecting telescopes some three metres in diameter, a number of special telescopes for studying the Sun, x-ray detectors, and a huge radio telescope. The instruments will be free of atmospheric interference. At present the 200-inch telescope on Mount Palomar can detect craters 3000 feet or more across on the moon. In space the same size of telescope will be able to distinguish craters only 300 feet in diameter.

It would seem that as a result of rockets and space travel a new era is opening up for the astronomer.

Life in the universe

One of the questions that has long intrigued Man is the possibility of there being living things in other parts of the universe. Could there be other living things like ourselves out there on other worlds in other planetary systems? The major difficulty that has got to be faced in investigational work of this kind is the immense distances that are involved. Leaving aside our own planetary system which will be dealt with in Chapter 2, our nearest neighbour, our nearest star, is four light years away. Since a light year is six million million miles then this star is twenty-four million million miles distant. Travelling at the speed of light

(186,000 miles per second) it would take a spacecraft four years to get there. To get to our nearest galaxy, Andromeda, travelling at the same speed, would take two million years.

We get an even better idea of the immensity of this journey when it is realized that *Mariner 2* in its voyage to Venus in 1962 averaged nineteen miles per second over its 180 million mile journey. Travelling at this speed it would take more than 39,000 years to get to the nearest star and more than 19,000 million years to get to Andromeda. Likewise it is not surprising that we on Earth have had no visitors from other worlds. Since, as we shall see, there is unlikely to be any life on the Sun's planets then distances involved in coming from other areas make this an impossibility, unless some form of travel completely unknown to us has been developed. Furthermore, why come to our world – there are so many millions upon millions to be looked at. I would think that even if the journey were possible then the chances of our world being chosen for visitation would be statistically almost non-existent. It looks as if exploring the universe outside the solar system will require not one, but generations of spacemen, unless of course we could beat time. It appears that even this is not outside the bounds of possibility. One conclusion of Einstein's Special Theory of Relativity is that time and space are intimately connected. This leads on to what has been called the 'clock paradox'. If you take a clock on an exceptionally high-speed journey in space then when you return with it to Earth it will record an earlier time than clocks left on Earth. If a man travelled say 500 light years out into space and then returned the 500 light year distance to Earth (all at the speed of light) he would find that Earth was 1000 years older whereas he himself would be only fractionally older than when he left.

Another major difficulty is concerned with the number of bodies to be searched. We have noted that there are 100,000 million stars in our galaxy, the Milky Way, and that there are 100,000 million galaxies in the universe. Thus there must be billions upon billions of stars presumably with planets and thus billions, upon billions, upon billions of planets in our universe. It has indeed been estimated that even within our own galaxy there are between 100 million and 100,000 million cold planets which could be considered suitable as potential supporters of life. Whether they have got life on them remains to be seen, but it would indeed be strange if our world, out of the billions upon billions of worlds in the universe, were to be the only one with life on it. Indeed it might be considered arrogant to assume so and much more reasonable to assume that the universe is teeming with life. An eminent American professor George Wald, has put it thus, 'We are not alone in the Universe and do not bear alone the whole burden of life and what comes of it. Life is a cosmic event. It has come many times, in many places. Should it fail here, all is not lost. Our kind will try again elsewhere.'

The only weapon that we have at our disposal in searching the universe for other living beings like ourselves is the radio telescope. As we shall see in Chapter 2, probes may be used within our solar system but beyond it and even in some cases within it, they have no value because of the distances involved.

It is less than 50 years since radio communication was invented, yet in that time we have achieved nearly technically perfect instruments. Many astronomers believe that within the next 50 years we should have them. A century is only a minute fraction of the age of our galaxy. Thus, it has been argued, on the galactic time scale a civilization passes abruptly from a state of no radio ability to one of perfect radio ability, so that if we examine a large number of life-bearing planets we should expect to find either a complete ignorance of radio techniques or a complete mastery of them. It is therefore quite logical to assume that those possessing this radio mastery will be capable of sending out powerful radio signals over long distances, as we can.

Physicists have calculated that these signals are most likely to be in the band between 1000 and 10,000 megacycles and that possibly frequencies around 1420 megacycles offer the best chance of success. Furthermore, they say, since we are attempting to intercept transmissions of great power where great range is presumably the goal we can expect the signals to be of narrow band width. This is very helpful as we will then be able to distinguish between such signals and those from naturally occurring sources as the latter are of broad band width. It was assumed the most likely system that any intelligent being would devise would be the simple sequence 1, 2, 3, 4 etc. given as a series of pulses indicating this sequence and repeated at regular intervals.

In January 1960 the Americans set up Project OZMA (named for the queen of the imaginary land of Oz – a place far away, difficult to reach and populated by exotic beings) using an 85-foot radio telescope to 'listen in' to the heavenly bodies. So far they have heard not even a whisper. Similar projects have been set up in USSR and in April 1965 some Russian scientists did report that a distant galaxy was giving out signals at regular 100 day cycles and that these were evidence of a 'super-civilization' in remotest space. Later however these reports were toned down and denied. The difficulties in this type of investigation are obvious – there are so many places in the universe to 'look' at and interpretation is such a difficult matter. Nevertheless it is still going on.

Further Reading

Sir Arthur Eddington 'The Milky Way and Beyond', *New Treasury of Science* pp 120–128. H. Shapley, S. Rapport and H. Wright (Eds), London 1965.

Frontiers in Astronomy: Readings from *Scientific American*, London 1970.

Larousse Encyclopedia of Astronomy, London 1966.
A. C. B. Lovell 'The Individual and the Universe', *BBC Reith Lectures 1958*, Oxford 1959.
Sir Bernard Lovell 'The Origin and Evolution of the Universe', *New Treasury of Science* pp 129–140.
G. Wald 'The Origins of Life', *Scientific American*, August 1954.

2. The Solar System

The origin of the solar system

The solar system consists of the Sun and its revolving planets including Earth. The entire system moves as a unit through interstellar space at a speed of about 45,000 miles per hour and this coherence makes it reasonable to assume that the bodies of the system not only share a common origin but also came into being at about the same period in time.

The first scientific attempt to explain the origin of the solar system came from the astronomer Laplace in Paris. He put forward the idea that at some remote period in time all of the material which now forms the solar system was a vast rotating nebula of gas. It cooled slowly and as it did so the gas contracted and as it contracted the rotation speeded up, eventually becoming so rapid that some of the gas broke away and formed a ring outside the nebula. This happened again and again to form more and more concentric rings. Finally the central portion condensed into the Sun and the material in each ring condensed to form the individual planets.

This theory was accepted for a long time but it was abandoned in the second half of the nineteenth century because of mathematical consideration which showed that rings of material successively detached from the periphery of a shrinking ring of gas would not join together to form single large planets but each ring would be transformed into a collection of small bodies like rings around the planet Saturn. It was also calculated that the speed of rotation of these rings must have been enormous and condensation at such speeds was an impossibility.

At the turn of the century two Americans, Chamberlin and Moulton, suggested that in the past our Sun was a star without planets. Then some 20,000 million years ago another star in its journey through space passed close to it – so close that great masses of gaseous matter were torn out of the Sun. Some of this material went into orbit round the Sun and eventually condensed into small fragments of material which joined up to form the planets.

Although this theory had many followers it was in turn undermined by another American, H. N. Russell, who in 1935 showed that for such an encounter to take place the approaching star would have to come so close to the Sun that the resulting planets would be in much closer orbit to it than they are now.

Many modern theories on the planetary system start with the Sun plus gaseous materials and dust. The most widely accepted version is that put forward by the Russian astronomer Otto Schmidt. It is close to the theory put forward by Laplace but it adds a new ingredient, namely that the circulating cloud consisted not of gas alone but a mixture of gas and dust from exploding supernovae, and that the planets were formed through a joining together of these particles. As the cloud rotated round the Sun the dust particles collided until appreciable aggregates formed with diameters of perhaps 100 miles. Then these aggregates, or 'embryos' as they have been called, began to collide with each other forming larger aggregates which were eventually to become the planets.

The planets are still being bombarded by dust particles. Earth for example is attacked by millions of tons of dust per year, but these are now stopped by our protective atmosphere of which more will be said later. We can detect these particles by means of satellites and of course occasional meteorites break through. These latter are believed to come from the region of asteroids between the planetary orbits of Mars and Jupiter where there are many thousands of minor planets or asteroids.

The Sun

Our Sun is a star and not a particularly big one or bright one at that. Like other stars it is a great globe of glowing gas. It is over 100 times the diameter and more than a million times the bulk of our planet Earth and it pours sunlight on to it at the rate of 4,690,000 horsepower per square mile of Earth surface. The equivalent of about 100 million Hiroshima-type atomic bombs reach the surface of the earth every day. This is however only a fraction of its prodigious output for it has been calculated that it puts out four million tons of radiant energy into space every second. We will leave it to the computer to calculate how much energy the Sun has given forth in all the years of its existence. So great is the radiation rate of the Sun's energy that it loses about four million tons in mass every second. Despite this rate of loss it is estimated that the Sun will survive as a source of energy for at least another 16,000 million years. The surface temperature is 6000°C but as we move inwards the temperature rises sharply so that at the core of the Sun the temperature is in the region of eighteen million degrees. Even at the surface, or rather just above and below it, there is however great variation. The sunspots which look like depressions in the solar surface and which are regions of intense electromagnetic energy are comparatively cool, being only 4000°C. On the other hand the chromosphere, which is the brilliant rim of light that can be seen easily at a total eclipse, has a temperature of about 20,000°C. The corona, the streams of light emanating from the sun sometimes known as the 'iron crown' and now known to be made up of elements such as iron, calcium and nickel, reaches a temperature of about a million degrees.

Like all other stars the Sun is gaseous throughout; the

gas is mainly hydrogen and this is the source of its energy. At high temperatures four hydrogen atoms combine to produce one helium atom. The helium atom is lighter than the sum of the four hydrogen atoms by 0.7% and this deficiency of mass is turned into energy which passes from the interior of the Sun to the surface in a steady flow. Although the Sun has been carrying out this reaction for thousands of millions of years it appears that it still consists almost entirely of hydrogen – enough to keep the process going for thousands of millions more years.

The description of the source of our power given by Miss Payne-Gaposchkin in her brilliant article on the Sun (see Further Reading) cannot be bettered. 'The Sun, indeed, holds a mirror to the cosmos. Like all other stars it is a globe of glowing gas, hotter and denser within. Its surface is a seething, surging sea of atoms. Plumes of gas float above it; glowing filaments surge upwards; shining fountains cascade downwards. Giant tornadoes swirl through the surface; spicules rise and dissipate like darting flames. Dazzling flares blaze up and vanish. A brilliant chromosphere rings it; and around it gleams the aura of the corona. Powerful magnetic forces play across the surface; atoms and electrons spray into space. As it spins on its axis the equator pulls steadily ahead; and across its face proceeds the slow rhythm of the sunspot cycle, waxing and waning every eleven years. The spectacle is impressive in itself. As the mirror of the cosmos it is stupendous. Other stars are doing the same things and these stellar habits are the clue to their history.'

The planets

The early astronomers differentiated between what they called the 'fixed stars' and the 'wandering stars'. We now know that the fixed stars are indeed stars but the 'wandering stars' are not stars at all but are the Sun's planets; with the unaided human eye the ancients could see five planets which in order of increasing distance from the Sun are Mercury, Venus, Mars, Jupiter and Saturn.

The invention of the telescope increased the number of bodies that could be seen in the heavens and in 1781 William Herschel announced his discovery of the planet Uranus which is so far away from the Sun that its finding immediately doubled the size of the solar system which until this time had been thought to have Saturn as its outermost planet.

Later, astronomers and mathematicians began to predict that there was yet another planet still further from the Sun than Uranus. They based their predictions on the fact that the path being followed by Uranus did not correspond to its theoretical orbit and they postulated that it was being disturbed by some unknown planet.

Two astronomers, Adams in England and Leverrier in France set out to find it. In August 1846 Leverrier after extensive calculations informed the Académie des Sciences in Paris that the new planet lay at a distance thirty times greater than the distance of the Earth from the Sun and he further predicted the exact position in the sky where it could be found. Three weeks later Galle of the Berlin Observatory found the planet in this position. Meanwhile Adams had reached the same conclusions, but he had some difficulty in persuading the Astronomer Royal that he was right so there was some delay in searching for it. Happily, although there was much acrimony at the time (though not between the two main actors), posterity has decided that both Leverrier and Adams should be given equal credit for their part in the discovery. And so Neptune, at nearly twice the distance of Uranus from the Sun, was added to the solar system.

In the 1920s for much the same reasons that led astronomers to the discovery of Neptune, it began to be suspected that there was yet another planet lying further out and its position in the sky was predicted. Finding the planet was however a much more difficult task because it was so far away, lost in the myriads of stars in the heavens. A systematic photographic search was begun by a young American, Clyde Tombaugh at Lowell Observatory, Flagstaff, Arizona and on 23 January 1930 he was rewarded by a faint spot on his photographic plate. This was the new planet, Pluto, and to date Pluto marks the outermost limit of our planetary system.

It is interesting that once these three planets, Uranus, Neptune and Pluto, had been discovered and charted, astronomers going over old records found that they had actually been charted in the past though they had been mistaken for stars; Uranus more than a dozen times between 1690 and its discovery in 1781, Neptune three times – twice in 1795 and one in 1845 – before its discovery in 1846, and Pluto four times – in 1919, 1921 and twice in 1927 before discovery in 1930.

Thus we have nine planets in the solar system – Mercury, Venus, Earth, Mars, Jupiter, Saturn, Uranus, Neptune and Pluto. They move round the Sun in what are to all intents and purposes concentric orbits. Mercury is closest to the Sun at 36 million miles, next comes Venus 67.2 million miles away and then Earth which is 92.9 million miles distant. Mars is further out at a distance of 141.6 million miles and then we take a big jump to Jupiter which is 483.3 million miles from the Sun. Saturn is nearly twice as far away again at 886.2 million miles, Uranus twice again at 1783 million miles. Neptune is 2794 million miles away and furthest out of all is Pluto at 3670 million miles.

The diameter of Earth is 7927 miles and very close to this is Venus with a diameter of 7700 miles. Both Mercury and Mars are much smaller having a diameter of 3100 and 4200 miles respectively. Then we come to the giant planets Jupiter and Saturn which at 88,800 miles and 74,200 miles have diameters approximately ten times that of Earth. Uranus and Neptune are rather similar to each other at 29,300 miles and 27,800 miles diameter respectively. The diameter of Pluto is not known for certain though it is thought to be very small. Since 1961 astronomers investigating the solar system have had a new weapon added to their armoury. It is echo-sounding or as it is better known, radar, which consists essentially of beaming radio waves on to targets (in this case the planets) and then building up a picture of the targets as a result of the echoes which are sent back. Using this device much information has been gained about Venus, Mercury, Mars. Radar has the great advantage of not being affected by the Earth's or the planet's atmospheres.

Asteroids

In addition to the planets there are some 1500 asteroids or minor planets so far discovered which are mostly contained between the orbits of Mars and Jupiter, although some are further out than this. It is certain that there are many more of these waiting to be discovered and some astronomers have predicted that there may indeed be between 50,000 and 100,000 of them.

They are very small, the diameter of the largest one known, Ceres, is 437 miles; Pallas is 287 miles; Vesta 240 miles and Juno 140 miles. These are the only asteroids that it has been possible to measure with any accuracy. Some of the others are no more than three or four miles in diameter.

The satellites

In addition most of the planets of the solar system appear to have satellites which revolve around them. Thus the Earth has one, the Moon. To this Moon must now be added the numerous 'artificial' man made satellites that permanently encircle her. Since the launching of the first artificial Earth satellite, *Sputnik 1*, by the Russians on 4 October 1957 more than 2000 artificial satellites have been placed in orbit for military, scientific and communication purposes. Mars has two satellites plus one 'artificial' one, Jupiter has twelve, Saturn has ten, Uranus has five and Neptune has two.

The diameter of the Moon is 2160 miles (just over a quarter that of Earth). Mars' two satellites Phobos and Deimos (the Greek name for Fear and Terror – the attendants of Mars) are ten and five miles respectively. The former was photographed in detail by *Mariner 9*. Only four of Jupiter's twelve satellites are of any great size. The largest of them are larger than the Moon and one of them, Ganymede, is as large as the planet Mercury. The largest of Saturn's ten satellites, Titan, is 3000 miles in diameter and the smallest, Phoebe, is 100 miles.

Saturn is unique in the solar system in that it has a series of concentric rings which are about 170,000 miles in diameter and situated in the plane of her equator. These can be seen with even a small telescope. There are three main rings. Ring A is greyish, ring B is very bright and the inner ring C is dark. It is now known that the rings are not either entirely solid or gaseous, since the former would be rapidly destroyed by the gravitational pull of the planet and the latter is not compatible with its observed opacity. The only reasonable explanation is that each ring is made up of millions of tiny particles so small and too close together to be seen individually.

The largest of Uranus' five satellites, Titania, is 700 miles in diameter; the smallest, Miranda, is 200 miles. The larger of Neptune's two satellites, Triton, is about 2300 miles in diameter. The other Nereid, is about 200 miles.

Thus our solar system is made up of the Sun, planets, satellites and asteroids spread over an area 7000 million miles wide. We should note however that we may not yet have determined the ultimate limits of our system and time will tell whether there are still more and farther planets lurking beyond our ken.

Life in the solar system

There has been considerable speculation as to the possibility of there being living things on the solar planets in addition to those found on Earth. Before going on to examine the conditions that are known to prevail on these planets let this be said. It is meaningless to consider the possibility of there being forms of life with attributes which are different from those known to us. This would merely take us into the meaningless land of speculation.

We must look for forms of life based on the carbon atom. There is no evidence that chemicals on other planets differ from the chemicals known on our planet and among them carbon is unique. Not only can it form long chain molecules – carbon joined to carbon joined to carbon and so on – but these carbon atoms can also form side chains with other carbons or with other elements to make up the intricate complex chemicals which we know are essential to living systems. Thus we look for life based on the carbon atom – life similar to our own.

Conditions necessary for living things

From our knowledge of the conditions under which living things grow on Earth we can list some of the conditions that must be satisfied before we can expect to find life on other planets. These are:

(1) Temperatures must not be too high or too low. The molecular arrangement of living matter is easily upset and temperatures of around 100°C will kill most higher forms of life and, although the spores (small particles of living material covered by a protective coat) of some bacteria will withstand this temperature, all of them will be killed on exposure for a short period to 150°C.

Low temperatures do not disorganize living systems but most forms of life cannot survive and those that do remain in a state of suspended animation as their energy-liberating enzyme systems cannot function at low temperatures. Many species of bacteria are however able to survive in extreme cold, as low as the temperature of liquid hydrogen (−250°C), provided they have also been dehydrated in the process of freezing, and then they will once again multiply in the presence of water and suitable nutrient when the temperature is raised to 25°C.

The seeds of some of the higher plants can withstand very low temperatures; for example seeds of the annual sunflower (native home USA), of *Convolvulus tricolor* (from southern Europe) and Balsam (from tropical Asia) and of other plants have been subjected for more than four days to the temperature of liquid air (−183°C to −192°C) and have germinated normally when returned to normal temperatures. Some seeds have been germinated after maintaining them for one to six hours at the temperature of liquid hydrogen (−250°C). These seeds included large types such as peas and marrow-pips as well as the very small seeds of *Mimulus moschatus*. This power to withstand very low temperature is due to the low water content of these seeds. If wetted they become very susceptible to such low temperatures, possibly due to the damage caused to their tissues by ice crystals.

(2) A supply of oxygen is necessary for all animal and most plant life. The oxygen is needed for the breakdown of food with the consequent yielding of energy which is used for the numerous activities of the living organism – moving, growing, reproducing and so on.

Some bacteria and many yeasts can live and grow in the absence of oxygen, and indeed to some species of bacteria the presence of oxygen is poisonous and they cannot survive for long in a normal atmosphere containing 20% oxygen. They are said to live anaerobically and they obtain their energy from foodstuffs not by using oxygen but by fermentation. This process does not however yield as much energy from a given unit of foodstuff as does the reaction using oxygen. This probably accounts for the fact that all higher forms of life, plant and animal, normally require oxygen though some – especially in the plant world – can survive for a short period by living anaerobically in the absence of oxygen.

(3) A supply of carbon dioxide is needed by most forms of plant life in order to build up foodstuffs such as sugars, and through them starches, fats and proteins, by the process known as photosynthesis. All animals require these foodstuffs that have been built up by plants.

(4) The presence of water either in the liquid form or as water vapour seems to be essential for both plant and animal life although, as we have noted above, some forms and some seeds can survive in the virtual absence of water. Some bacteria, algae, lichens and mosses can survive in a dry state for many years and they will grow again when moisture is provided.

(5) Most gases such as ammonia, chlorine, carbon monoxide and hydrogen sulphide are poisonous to most forms of life; although their presence in an atmosphere would not prove conclusively that there could be no life there, it would be strong presumptive evidence.

Could there be life on the planets?

Let us now look at some of the planets to see if any of them could possibly support life.

Nearest to the Sun is Mercury. Because of the way in which it moves round the Sun only a little more than 70% of its globe is visible from the Earth, the other 30% remaining in eternal darkness. The surface temperature is about 410°C. Water therefore could not exist even as a vapour in the thin atmosphere which may surround the planet and because of its low mass it is very unlikely that Mercury retains any atmosphere at all. Some of the surface markings on the planet become obscure from time to time as if covered by a whitish veil and although early observers felt that these veils might be evidence of water vapour it is now considered that they are probably fine dust clouds. In March 1974 the spacecraft *Mariner 10* passing within 21,700 miles sent back pictures showing that its surface is heavily cratered and covered by low hills. Because of its high temperature and its lack of water it is impossible for any form of life to exist on Mercury.

Of all the planets Venus most nearly resembles the Earth in size and mass. At first sight indeed Venus and the Earth are almost perfect twins. But appearances are deceptive. On the 14 December 1962 the Americans sent a space probe (*Mariner 2*) on a 180 million miles journey to Venus. The trip took 109 days and the probe passed within 21,648 miles of the surface of the planet. This and other instrumented flights such as *Mariner 5* by the Americans and the Russian atmospheric probes *Venera 4, 5, 6* and *7* have indicated that the surface has a temperature of about 485°C, hot enough to melt lead, and it could not possibly sustain life. Apparently there is also very little water vapour in the atmosphere – less than a thousandth of the Earth's atmosphere. The dense atmosphere is made up mostly of carbon dioxide with a little nitrogen and a trace of oxygen and the surface pressure is about 100 times as great as Earth's atmospheric pressure. A Russian spacecraft reached Venus on 22 July 1972 and transmitted information for 50 minutes.

Mars is the fourth planet from the Sun. It is sometimes known as the red planet and no doubt this redness suggested its identification with the great Roman god of war. Under good conditions a great deal of detail can be observed on the surface of Mars. Some three-fifths of the surface is reddish in colour and this is probably due to iron oxide. Standing out clearly in contrast to this are dark markings which are grey or slightly green. Another very characteristic feature is the two white patches which occupy the polar regions of the planet. At one time it was thought that these caps were made up of ice but it is now fairly certain that they consist of frozen carbon dioxide and are very thin.

Mars has very little atmosphere and it consists chiefly

of carbon dioxide. There is no appreciable amount of water and it appears that unlike Earth there is no nitrogen. The average surface temperature is well below freezing point although there is considerable variation. In the Martian tropics the highest temperature may be well above zero reaching 10-30°C. Toward evening the temperature drops sharply and at sunset is down to zero. During the night there is a further fall down to −40°C and at sunrise the recorded temperatures are −20°C to −30°C. These great variations in temperature are proof of the thinness of the atmosphere and of the almost complete absence of water. The surface is being constantly bombarded by radiation from the Sun creating conditions which would make life impossible.

In 1877 the astronomer Schiaparelli from the Milan Observatory reported a system of lines linking one dark spot to another covering the surface of the planet like a network. These were given the name 'canals' and have been so known ever since. This is an unfortunate title because from there it was but a short step to proposing that these canals were made by intelligent beings on the planet. The American *Mariner 4* spacecraft in 1965 after its 287 million miles, 167 day journey sent the first television picture of Mars which showed the surface to be covered by circular craters very much like the Moon with little trace of these straight lines or canals. Four years later the pictures sent back by *Mariner 6* and *Mariner 7* showed that the surface was not uniformly cratered but that large areas of it were rather chaotic with large volcanoes and canyons. These pictures have been confirmed by the robot *Mariner 9* which was put into orbit around Mars in November 1971. *Mariner 9* thus became, by the way, the first man made satellite of another planet. In its orbit it came at its nearest point to within 1000 miles of the surface and 10,000 miles at its most distant.

Some astronomers have interpreted its findings as indicating that Mars is just beginning to 'boil' inside, going through a phase similar to one that the Earth went through early in its history. They believe also that the thin atmosphere may just be the beginning of a 'real' atmosphere.

The Russians landed a capsule, *Mars 3*, on Mars in 1971 but unfortunately soon after landing it ceased to transmit. We eagerly await the arrival on Mars of the four Russian space probes now on their way. The Russians may find it possible to obtain samples of material from Mars around 1980 and manned landing by either the USA or the Russians is possible in the 1980s.

Compared with the other members of the solar system Jupiter is a giant, being 11.2 times the diameter and more than 1300 times the volume of the Earth. It is however surprisingly light in density its mass being only 318 times that of Earth. The force of gravity is such that a 12 stone space traveller would find that his weight was nearly 32 stone on Jupiter.

Because of its great size Jupiter is a brilliant object in the sky and one of its most characteristic features is a series of belts encircling the globe parallel to the equator. They can be readily seen by telescope and they consist of atmospheric not surface features. They are of different widths and shading and variations in colour appear ranging from pink to purplish brown. Another marking is the Great Red Spot which is about 30,000 miles long and 7000 miles wide. Its nature is unknown but it drifts in longitude, having been observed to drift almost 44,000 miles in fourteen years. There are times when it loses its reddish colour and becomes a greenish white which is only just visible.

The planet has got a very thick atmosphere which contains large quantities of methane and ammonia, and hydrogen and helium are probably also present. The most recent theories picture Jupiter as consisting of a large inner cave of ice (under immense pressure) surrounded by a vast ocean of compressed and liquified hydrogen and helium merging into an atmosphere of hydrogen. In this atmosphere clouds of methane and ammonia, violently churned, form the belts and spots visible from the Earth. Because of its immense distance from the Sun the temperature of the planet is in the region of −130° C to −155° C. *Pioneer 10* reached its closest point to Jupiter on 4 December 1973 after a journey of nearly two years and a stream of colour pictures were sent back to Earth. It is the first probe to penetrate beyond Mars and it is now speeding on its journey out of the solar system. In the hope that it may be picked up by intelligent beings, Man has included his 'visiting card' – an engraved plaque showing where the probe has come from and including outline drawings of a man and a woman.

After Jupiter, Saturn is the second largest planet in the solar system being 9.4 times the diameter and 744 times the volume of the Earth. Like Jupiter it is of very low density and the surface gravity is probably only about one fifth more than it is on Earth. We are never closer to Saturn than 750 million miles so, in spite of its great size, it never appears any more than a small disc. As mentioned earlier, one of its distinctive features is the series of concentric rings which are probably made up of fine debris. Like Jupiter it has a series of belts running parallel to the equator but they do not stand out quite so clearly as those on Jupiter. The bright equatorial region is pinkish in colour while the belts are brownish grey. The polar regions are bluish. The surface is probably extremely unstable being made up of gas which liquifies towards the inside of the planet. The temperature has been estimated as approximately −150°C and there is both ammonia and methane in the atmosphere. Because of the very low temperature, however, it is thought that the ammonia must be in solid form and that certain large white spots which are sometimes seen in the equatorial region might be 'ammonia snowstorms'. As in the case of Jupiter we cannot even imagine what the outer layers of the surface of Saturn are like except that they are quite unlike anything here on Earth. Saturn may be investigated by probes in 1977.

It is of interest to note that one of Saturn's satellites – Titan – possesses an atmosphere which is very similar in composition to that of Saturn itself. This is the only recorded case of a satellite having an atmosphere of any importance.

Although classed among the giants of the solar system, Uranus is much smaller than either Jupiter or Saturn. Its diameter is about four times, but its density only about a quarter that of Earth, but being so very far away it is practically invisible to the naked eye. When viewed through a good telescope it appears to be greenish-blue in colour. In constitution it is probably very similar to Neptune and Jupiter having an atmosphere made up of a thick cloud of methane and above this an atmosphere of hydrogen. The temperature is about –160°C so that ammonia if present must be in the solid state.

Stratoscope II, a telescope 36 inches in diameter borne aloft in a balloon in January 1973, has obtained the first high resolution pictures of the planet at an altitude of 24 kilometres. This is above the altitude where our atmosphere interferes with the view.

Neptune is very similar to Uranus in size having a diameter of 27,800 miles against Uranus' 29,300. In mass it is about seventeen times that of the Earth. Because of the great distance from the Sun it has a very low illumination and because of its great distance from the Earth it is quite invisible to the naked eye. A good telescope reveals that it is of a greenish-blue colour, rather similar to that of Uranus but little detail can be made out. It appears to have dark belts running parallel to the equator like those of Saturn, Jupiter, Uranus and Neptune. It has also got a similar atmosphere made up largely of methane. Because of the low temperature –170°C much of the methane gas has probably turned to liquid, whilst it is said that at this temperature most other gases will have solidified.

Very little is known about Pluto whose orbit, as we have noted, forms the furthest boundary of the solar system. It is such a faint object that a very large telescope is required to see it at all. Its diameter cannot be measured directly though it has been estimated to be about half that of Earth. It has about one tenth of Earth's volume although its mass is probably about the same as Earth's suggesting that it may be about ten times as dense. All of the above is however uncertain and at best a good guess. It has also been suggested that Pluto is covered with liquified or frozen atmosphere smooth enough to act as a sort of mirror. There are some who, because of its orbit, believe that Pluto is an escaped satellite of Neptune. Because of its small size Pluto belongs to the group of terrestrial planets – Mercury, Venus, Earth and Mars – rather than the gaseous planets – Jupiter, Saturn, Uranus and Neptune.

It is possible that the whole solar system will have been explored by automatic probes by the end of the twentieth century.

Meteorites

Apart from probes and the spectroscopic examination of planets to determine if conditions might be suitable for living things there is yet another way of looking for life beyond planet Earth. It is by the examination of meteorites.

The objects known as meteors are well known. These are seen as brilliant flashes of light, 'shooting stars', that pass across the night sky. They are caused by materials from outer space which in the course of passing through our atmosphere are heated to a brilliant incandescence. They cannot be seen so long as they are beyond the Earth's atmosphere and they become visible only when they enter it. Generally these bodies are burned out before they reach the surface of the Earth but some survive and land on the earth as lumps of stone or metal. Such survivors are known as meteorites.

The fall of a meteorite is a relatively rare event since in most cases they are vaporized before they reach the Earth's surface. It has been estimated that those weighing say 5 tons in space and yielding a meteorite of 1000 lbs on Earth occur at the rate of about two every 3 years over the area of the United States. Estimates have been made of the rate of occurrence of really large ones, e.g. 50 tons in space yielding a meteorite of a few tons on arrival, and it has been determined that these occur at the rate of about one every 30 years over the whole of our globe.

Some enormous meteorites have been recorded. For example the Meteor Crater in the Arizona Desert is a roughly circular pit 4200 feet in diameter and about 570 feet deep and must have been formed by an enormous meteorite which may lie buried in fragments hundreds of feet below the surface.

One fall that occurred on 30 June 1908 in Siberia was described by onlookers as 'a ball of fire across the sky' followed by an enormous crash. Explosions were heard up to a distance of 600 miles away. At distances of 100 to 150 miles men and horses were blown over. Enormous waves formed on the rivers, houses collapsed. At a distance of 250 miles from the point of impact a jet of flames was observed which must have been 12 miles in height. The remains of the meteorite itself have never been found but it has been estimated that its total mass could not have been less than 40,000 tons. The largest meteoric mass so far discovered, in French West Africa, is estimated to weigh a million tons.

The composition of meteorites is varied but no new or unknown chemical element has been discovered. The most frequently occurring elements are iron, oxygen, nickel, silicon, and magnesium. Some also contain various gases, such as hydrocarbons, carbon monoxide, nitrogen and hydrogen. Their origin is however still a mystery.

In March 1961 three scientists, Nagy, Hennessy and Meinschein, analysed a sample of a meteorite which landed at Orgueil in France in 1864. It belongs to a special class known as the 'charcoal meteorites' because they are composed of friable material which breaks up readily. They are rather rare.

The scientists reported that the volatile portion of the organic matter that was present consisted largely of the group of chemicals known as saturated hydrocarbons which resembled the hydrocarbons in the products of living things on Earth and they suggested that they had probably been made by living systems on other worlds. The chemicals were present in such amounts as to rule out the possibility of their being contaminants picked up by the meteorite after it entered the Earth's atmosphere.

There were questions however as to whether they had been formed by living processes. Some scientists suggested that such organic materials might have been the forerunners of living things both in other worlds and here on Earth. As we shall see in the following chapters there is fairly strong suggestive evidence that living things did come into being via organic chemicals formed here on Earth.

Nagy and his workers went even further and announced in November 1961 that they had discovered 'microscopic sized particles' resembling fossil algae in relatively large quantities in some of their meteorites. Some of these reacted to biological dyes which are used to determine the presence of nucleic acids which are vital to living things.

The presence of these materials was not questioned. What was questioned was their origin and serious doubt was thrown on their extra terrestrial origin when one of these organized elements which had been photographed and published by Nagy was shown by another worker to be ragweed (*Ambrosia elatior*) pollen – presumably picked up on Earth.

The claim by Dr F. D. Sister of the US Geological Survey to have grown living microbes from these meteorites was not questioned either but again the possibility of any living organism surviving the intense heat of the meteor's passage through our atmosphere is next to impossible and it is generally agreed that these microbes grown by Sister were contaminants picked up by the meteorites on Earth.

We are left then after this rather brief survey of our universe with the conclusion that no living things have yet been detected beyond the planet Earth. Furthermore the possibility is remote that there are any living things on any of the planets or other bodies of the solar system but that on statistical grounds there is good reason for believing that the universe might be teeming with life. One planet that is certainly teeming with life is Earth and we shall go on to examine it in our next chapter.

Further Reading

G. Abell *Exploration of the Universe*, 1969.

A. F. O. Alexander *The Planet Saturn*, London 1962.

B. J. and P. F. Bok *The Milky Way*, London and Cambridge (Mass) 1957.

I. O. Evans *The Earth*, London 1970.

Frontiers in Astronomy: Readings from *Scientific American*, London 1970.

M. Grosser *The Discovery of Neptune*, London and Cambridge (Mass) 1962.

F. Heide *Meteorites*, London and Toronto 1964.

J. N. James 'The Voyage of Mariner II', *New Treasury of Science* pp 171–187.

R. Jastrow and A. G. W. Cameron (Eds) *Origin of the Solar System*, New York and London 1963.

Larousse Encyclopedia of Astronomy.

Peter Millman 'Our Neighbour World', *New Treasury of Science* pp 92–102.

Iain Nicolson *Astronomy*, London 1972.

Cecilia Payne-Gaposchkin 'The Sun', *New Treasury of Science* pp 103–113.

B. M. Peek *The Planet Jupiter*, London 1958.

G. D. Roth *The System of Minor Planets*, London 1962.

W. Sandner *The Planet Mercury*, London 1963.

M. Schwarzschild *Sun and Stars*, 1958.

3. The Planet Earth

The Earth – some statistics
Earth, the third planet in the solar system, lies between Venus and Mars. It is roughly ellipsoidal in shape, the slight bulge at the equator being approximately what we would expect to find in a rotating fluid. This equatorial region has bulged out under the force of the spinning earth and the polar regions have become proportionately flatter. The total volume of the Earth is 260,000 million cubic miles and it weighs 5887 trillion tons. At the equator the circumference is 24,902 miles. It makes one complete revolution round the Sun every 365 days, 5 hours, 48 minutes 46 seconds and one complete rotation on its axis in 23 hours 56 minutes at a speed of more than 1000 miles per hour at the equator. Further north and south the rate is slower, e.g. at the latitude of the United Kingdom the speed is about 700 miles per hour.

Formation of the Earth
In common with the other planets of the solar system it seems clear than planet Earth was formed by an accretionary process as dust particles collided and joined together. The process is still continuing and Earth picks up about fourteen tons of cosmic dust each year. The original Earth was not a molten mass as was once thought. It was a cold body which later became hot and probably molten due to the activity of radioactive chemicals trapped inside it.

Radioactivity
Radioactivity consists in the transformation of one ele-

ment into another with a consequent release of energy. For example uranium is changed through a series of decay processes into radium and then by further decay processes into lead, energy being released at every step. In a similar way potassium decays to form argon and rubidium decays to form strontium. These radioactive decay processes are not affected in any way by the environment. They continue at a set unchanging pace, no matter what other changes are taking place around them. This means that we have a steady and reliable clock to measure past ages. The lead that is produced from uranium can be readily identified as being different from the lead that occurs naturally and so if we find a rock containing both uranium and this 'special' lead, knowing the rate of decay of uranium, we can then calculate how long it has taken the uranium to produce the lead deposit.

Using this method geologists have been able to make a fairly accurate guess as to the age of the Earth. Recent estimates suggest that it is in the region of 5000 million years old. Most of the heat flowing out of the Earth is produced by the radioactivity of uranium and thorium. These radioactive materials are probably strongly concentrated in the outermost layer of the Earth. If they were deeper in then much of the heat would be trapped and the Earth would melt.

Inside the Earth

Our knowledge of the inside of Earth is still hazy and much of the information that we have got comes from a rather surprising source – earthquakes.

Each year our planet is shaken by ten or more major earthquakes and even the smallest of these releases about a thousand times more energy than an atomic bomb. The earthquake causes shock waves which travel through the Earth and the measurement of the path of these waves give us a picture of the Earth's interior.

There are three main types of shock waves: primary

waves which are compression and expansion waves rather like those of sound. They travel through both the solid and liquid parts of the Earth; secondary waves which vibrate at right angles to the direction of travel as light waves do. These waves can pass through solids but cannot pass through liquids; surface waves which appear in the upper twenty miles or so of the Earth's surface radiate out rather like the ripples on a pond when one throws a pebble into the centre.

The speed of travel of these various waves varies with the depths from which they are coming, e.g. primary waves travel at their maximum speed – 8½ miles per second at a depth of 1800 miles and at about 3 miles per second in rocks nearer the Earth's surface. Secondary waves travel at about two-thirds the speed of primary waves. Both primary and secondary waves are deflected when they encounter boundary layers within the Earth.

Scientists have made great use of all this information to construct models of the interior of the Earth. There is a large central core which has a diameter of about 4320 miles and whose boundary lies about 1800 miles below the surface. This core is not uniform. The innermost part which has a diameter of about 1600 miles is solid, twice as rigid as steel, and is probably made up of iron and nickel. The outermost part of the core, rather surprisingly, appears to be liquid. This has been long assumed to be made up of iron and nickel too but suggestions have been put forward that it may be compressed hydrogen.

All of the Earth outside the core is termed the mantle and this mantle (except for the oceans and volcanic patches) is solid. This is shown by the fact that both primary and secondary shock waves pass through it. It is not however completely uniform because there is an important boundary just twenty miles below the surface of the earth. From this boundary to the surface is known as the crust.

This twenty miles of crust is made up of three kinds of rocks. On the outermost surface there are the sedimentary rocks such as the sandstones, limestones, shales and clays which are the hardened, compressed remains of sediments laid down by ancient rivers, seas and glaciers. They may be flat as in the great plains or thrown into great convoluted folds as in our mountain chains. Beneath the sedimentary rocks lie the metamorphic rocks. These are sedimentary rocks that have been 'metamorphosed' or altered by intense heat and pressure. They include the schists, gneisses, quartzites and the slates. Beneath the metamorphic rocks and extending right down to the boundary layer are the plutonic or igneous rocks made up largely of granite. It is not known how these igneous rocks have been formed but it is thought likely that they have crystallized out from molten rock deep in the crust.

Pressures and temperatures

The densities, pressures and temperatures that appear to be present in the Earth are rather staggering. For example, some 200 miles down in the Earth the pressure is about 100,000 atmospheres and at the junction of the mantle and the core the pressure is $1\frac{1}{3}$ million atmospheres. Within the core it is probably about 4 million atmospheres. These pressures can be compared with the pressure at sea level which is 1 atmosphere.

The measurements of temperature are less certain than those of pressure. As one goes down a deep mine the temperature rises 30°C for every mile in depth. Judged on this basis, this would mean that the temperature at the centre of the earth would be 100,000°C. It is now thought however that there is not a steady rise with mileage and a figure of about 6000°C is a more reasonable estimate.

The shape of the surface

The surface of the Earth is very irregular, ranging from Mount Everest on the border between Nepal and Tibet, which is 29,028 feet above sea level to the depths of the

Marianas Trench off the Philippines in the Pacific Ocean which goes down to 36,198ft below sea level.

The problem of the formation of the mountains remains one of the great unsolved questions of physical science. A number of theories have been put forward suggesting that various agencies have been at work. Among these is the idea that contraction has probably played an important part. Just as an apple develops wrinkles on its skin as it dries out so it is suggested the Earth's crust wrinkled in a similar way as the cooling process set in, following in the wake of intense heating by radioactivity. Furthermore, if the mantle was at one time a viscous fluid, then the heat emanating from below would set up convection currents leading to 'swellings' appearing on the Earth's surface. With changes in temperature there would also be profound changes in the crystalline structure of the minerals which could lead to large vertical displacements. Localized melting of certain minerals would also be very important so that the lighter components might gradually make their way through the overlying layers to the top and accumulate on the surface. On the whole we find that the lighter compounds are more concentrated in the crust, there being fairly high concentrations of compounds of sodium, magnesium, aluminium and calcium.

It seems likely that as the Earth cooled so the water, which had originally evaporated off to form clouds, began to fall until the whole planet became ocean. Beneath the water the continents began to rise due to stresses and strains such as those described above. Even today more than 70% of the Earth's surface is covered with water. The Pacific Ocean alone covers an area of almost 64 million square miles. In total, the water surface of the Earth is nearly 140 million square miles compared with a total land surface of 57½ million square miles. This land surface is made up of seven continents – Africa, Antarctica, Asia, Australasia, Europe, North America and South America. It appears however that the land masses were not always as they are now.

Continental drift

In 1620 Francis Bacon discussed the possibility that the western hemisphere had once been joined to Europe and Africa and in 1858 Antonio Snider noticing that the fossils from the same period in both America and Europe were very similar proposed that all the continents of the world had, at one time in the past, formed a single land mass.

Towards the end of the nineteenth century an Austrian geologist, Eduard Suess, noted that all the geological formations of the southern hemisphere were very similar to each other and he proposed that at one time South America, Africa and Australia had fitted closely together to form one mass for which he proposed the name Gondwanaland. By this time the fitting together of continents had become a geological exercise and in 1910 a German, Alfred Wegener, put forward proposals suggesting how the various continents had drifted apart by lateral displacement of the Earth's crust. He suggested that all the present day continents had been at one time joined together to form a super-continent which he called Pangaea. This raised considerable discussion but the general consensus of opinion among geologists was that it was nonsense. Wegener died in 1930 with his views unaccepted.

It was not until after 1945 that geologists began to look again at Wegener's work and to investigate with sophisticated instruments the theories that he put forward. Today as a result of evidence from many sources his theory is almost universally accepted and the theory of the evolution of continents by drifting has become as exciting an idea as Darwin's *Origin of Species.* Harry Hess and Robert Dietz in the United States suggested that great rifts appeared in the sea floor and through these rifts material from the Earth's interior welled forth, solidifying, spreading the sea floor, thus carrying or pushing the continents ahead of them. This created considerable interest and evidence of such movement came from the study of magnetic fields – these are the magnetic lines of force that run between the north and south poles. They magnetize the rocks that they pass through and thus leave a permanent record of their passage. Studies of these magnetic fields indicated that continental drift must have taken place. The poles in England's rocks could be precisely matched to those of North America only by closing up the Atlantic.

Further evidence from study of the Earth's magnetic field indicated quite clearly that the Atlantic Ocean was opening at the rate of one to two inches per year. Further evidence has come from computer studies on the 'fit' of various continents comparing not today's coastlines but the continental slopes in the oceans. The fit is quite remarkably good. The rock strata of the west coast of Africa have also been compared with the east coast of South America. The matching of the rock is quite remarkable, even including gold and tin layers. Furthermore fossil finds in Antarctica indicate that reptiles found there are identical with those found in Africa, India and China indicating that these continents and countries were at one time joined together.

Taking all of this evidence together it seems to be indisputable that Pangaea existed. Also some scientists as a result of all the evidence that has been accumulating have been able to describe what happened. The following quote is from a fascinating article by Samuel Matthews in the *National Geographic* of January 1973, 'North America and Africa separated 180 to 200 million years ago', says Walter C. Pitman II. 'A rift between Africa and South America appeared about 135 million years ago. Finally North America parted company with Europe only about 80 million years ago.'

'Pangaea split into two blocks – Laurasia in the north (North America, Europe and Asia). In the south was Gondwanaland (Africa, South America, Antarctica, Australia and India). Then South America began sliding westward. Africa parted from Antarctica. India broke free and sailed 5000 miles north. It collided with Asia only some 40 million years ago. The collision uplifted the high Tibetan Plateau and raised the Himalayas.'

The sea-going drill tower *Glomar Challenger*, drilling rock and mud, has proved beyond reasonable doubt that Africa, moving northwards, is closing the Mediterranean Sea. But just as seas and oceans are being closed so new ones are being opened up. It has also been calculated that the site of Los Angeles is moving towards that of San Francisco at the rate of two inches per year. The sites, at present 350 miles apart, will meet ten million years from now. Geologists as yet do not know what causes continents to move. The solution of this problem is going to be one of the great discoveries of this century.

The atmosphere

The atmosphere is the envelope of gas around the Earth. At the Earth's surface it is made up of 78% nitrogen and 21% oxygen, with the remainder being made up of traces of carbon dioxide and hydrogen together with some of the heavy gases such as argon, neon, krypton and helium. The amounts of water vapour vary from area to area and, indeed, above 10 miles, water vapour is practically absent.

The gaseous constituents remain fairly constant up to a height of 44 miles while traces of air have been detected well over 600 miles up. At about 10 miles up, due to the action of ultra-violet light on oxygen, a layer of ozone is formed. As we shall see in the next chapter this ozone layer is very important in protecting living things from the harmful effects of ultra-violet rays from the Sun.

The atmosphere is made up of four layers:

(1) The troposphere. This is the lowest main layer extending in temperate regions to about 7 miles up; in polar regions to 4–5 miles and in equatorial regions to 10–12 miles. The temperature falls with height by an average 10°C per mile (with small variations). The level at which the temperature ceases to drop and becomes stable is called the tropopause and here the temperature varies from −80°C near the equator to −40°C to −60°C near the poles. Most weather is generated in the troposphere.

(2) Above the tropopause is the stratosphere which extends upwards for about 30 miles. This is a very stable region where the temperature remains fairly constant. At the uppermost part of the stratosphere (the stratopause) the temperature is about 0°C.

(3) Upwards from the stratopause is the mesosphere where the temperature decreases again with height to a minimum of −90°C at about 50 miles.

(4) Above this cold layer is the thermosphere in which the temperature increases with height and reaches over 1000°C before merging with the rarified matter of interplanetary space. The thermosphere is included in the ionosphere which extends upwards for about 300 miles. The ionosphere is the region where radiation from the sun partially ionizes the gases, i.e. it strips off the outer electrons from their molecules so that the molecules become positively charged. This layer reflects radio waves of short wavelengths and so makes long distance radio communication on Earth possible.

Clouds

As we have noted, the amount of water vapour in the atmosphere is variable. It is derived from the evaporation of the seas, rivers and lakes of the world and its concentration depends on the temperature of the air. The higher the temperature then the more water vapour the air can hold. Water vapour is invisible but when the temperature falls below a certain level then it condenses around dust or other tiny particles in the atmosphere and becomes visible as ice crystals or droplets of water. Clouds are made up of such crystals and droplets. The form that the cloud takes depends upon the altitude (and therefore the temperature) at which it is formed. Thin filaments of cirrus, made up of tiny crystals, occur at an altitude of some 30,000 feet. At 3000 feet the clouds are of the dark nimbus (the rain clouds) and cumulus (cotton wool) types. Intermediate types occur between these levels.

Clouds serve not only as reservoirs of water but they also provide a protective blanket. They shield the Earth from the intense rays of the Sun and they prevent excessive heat loss from the Earth's surface.

According to Dr Helmut Landsberg, who is Director of Geophysics at the Air Force Cambridge Research Centre, the atmosphere has not always been as it is now. Not only has it undergone dramatic changes in the past but it is still changing. There is much less hydrogen and heavy gases in the Earth's atmosphere than on other planets; oxygen, nitrogen, carbon dioxide and water vapour occur in much greater amounts. It would seem that Earth must have lost much of its heavy gases at some stage in its evolution.

In order to do so its temperature must have been higher than now; it must indeed have been in the region of 14,500°F. At this temperature all of the other gases such as carbon dioxide and oxygen would escape too so that the Earth would lose all of its atmosphere. As the molten rocks cooled, however, a new atmosphere would be formed by gases such as carbon dioxide, nitrogen and water vapour, which had been dissolved within the rocks, being released. At this stage the surface temperature would be several hundred degrees Fahrenheit. According to Dr Landsberg, this atmosphere would be rather similar to the conditions found around present-day active volcanoes.

One point worthy of note is that in this early atmosphere there was no free oxygen. Where did it subsequently come from? A little would come from water when high temperatures would split some of the molecules into hydrogen and oxygen but this would probably be promptly taken up by the rocks in oxidizing processes akin to the rusting of iron. As we shall see in the next chapter it is now generally accepted that almost all of the oxygen in our present day atmosphere has come from plant-life. When there were no plants there was little or no oxygen.

Let us follow our primitive atmosphere as further cooling took place. At around 700°F, liquid water could exist simultaneously with vapour. With further cooling, more liquid water would be formed from vapour and the rains would begin. These torrential rains probably fell for centuries and formed the oceans of the world.

Dr Landsberg also discusses why our Earth is now able to retain its present atmosphere. Why do the gases not go streaming off into space as in the past? Two factors control the escape of gases from Earth. One is the temperature and density of the gases. The other is the gravitational pull of the Earth. In order to overcome this pull and escape into space we know that a body shot upwards must go faster than 7 miles per second. The average velocity for hydrogen in the Earth's atmosphere is 1¼ miles per second; for nitrogen and oxygen ¼ mile per second. The speed and direction of the molecules making up the gases is determined by temperature and density. At a temperature of 2370°F (which is probably the temperature 185 miles above the earth) it would take about 4000 years for most of the hydrogen to escape; but for nitrogen it would take 10^{45} years and for oxygen 10^{51}. It looks as if they are both going to be around for a very long time to come! It is of interest perhaps to note that the highest temperature recorded on Earth in modern times is 136.4°F at Aziza in Libya. The lowest is −125.3°F at Vostok in Antarctica.

The Moon

In July 1969 Man landed on the Moon. *Apollo 11* traversed the 238,900 miles from Cape Kennedy and the two American astronauts Neil Armstrong and Edwin Aldrin became the first men ever to step on land other than Earth. The dream of centuries was realized and Earth's nearest neighbour could now be subjected to detailed analysis. Proof had replaced speculation.

The moon landing was the culmination of a process that began in 1957 when the first artificial satellites were placed in orbit round the Earth and in those eleven years the pace of development has been breathtaking. On 13 September 1959 the Russian *Lunik 2* crash landed on the Mare Imbrium and a month later *Lunik 3* gave us a view of the far side of the moon – a side that had remained hidden from Man since the beginnings of time. Since 1957 until 1969 no less than 34 spacecraft – American *Rangers*, *Surveyors* and *Orbiters* and Russian *Lunas* – have followed in the wake of the first. Seven have made soft landings.

Since 1969 there have been manned landings by the Americans and in November 1969 the Russian *Luna 17* delivered to the surface of the moon a remote-controlled vehicle *Lunokhod* which drove around the lunar surface for a few months, radioing photographs and scientific observations back to Earth.

As a result of the satellite programme the entire surface of the Moon – front and back – totalling 38 million square kilómetres in area, has been photographed in considerable detail so that together with the information brought back from the manned landings and sent from *Lunokhod* we now know nearly as much about the surface of the Moon as we know about the surface of the Earth.

Some statistics

The Moon has a diameter about a quarter that of the Earth and a surface area less than half that of the Atlantic Ocean. Its gravitational pull is about one sixth that of Earth's. The rock samples that have been brought back indicate that it is about the same age as the Earth and the surface materials are similar to volcanic basalt. Unlike the centre of the Earth however the interior of the Moon seems to be cold and solid.

The origin of the Moon

The origin of the Moon is a problem that has long intrigued scientists. Was it once part of the Earth until it was torn out of the Pacific Ocean? Were both Earth and Moon formed simultaneously from one large body? Was the Moon captured by the Earth as a satellite?

Before looking at possible answers to the questions it is of interest to note that the Earth-Moon relationship is rather exceptional in the solar system with regard to size. The Moon is about one eightieth the mass of Earth whereas Ganymede, which is one of Jupiter's satellites and is twice the mass of the Moon, is less than one ten-thousandth that of Jupiter; and Titan, similar in mass to Ganymede, is only about one four-thousandths that of its 'parent' Saturn. The Earth-Moon system has sometimes indeed been described as a 'double planet' rather than a satellite system.

Returning to origins, it appears that it is dynamically impossible for the Moon to have been ejected from Earth but the theory that they may both have developed simultaneously from one massive body is at least a possibility. Dr R. A. Lyttleton of Cambridge goes further and has suggested that the Earth, Moon and Mars may all have been formed from this one body. He visualizes that this body might have elongated to become roughly egg-shaped. Swellings would then appear at either end to give an unequal dumb-bell with the larger swelling being the forerunner of the Earth and the smaller the forerunner of Mars. Before they separated, however, a third swelling would appear on the 'stalk' of the dumb-bell and this swelling would be the forerunner of the Moon. The three swellings would then separate. Mars would 'escape' to pursue its own orbit but the Moon would be captured by Earth's gravity and would become its satellite. The theory is a very attractive one and it is dynamically acceptable.

The third theory, that the Earth captured a once independent small satellite, the Moon, is also acceptable provided one invokes the help of the Sun. It appears that direct capture alone by the Earth is not possible because the energy required to bring one body into close proximity to another would also carry it away into infinity or alternatively cause the approaching body to be absorbed by the Earth. Some additional body is required to 'damp down' this energy and this role can be played by the Sun. It seems that, without the presence of the Sun, the Moon could and would escape quite readily from the Earth.

The Lunar environment

As a result of the numerous observations that have been made, the United States Air Force has produced a series of charts which map the Moon in great detail at sixteen miles to one inch. There are great smooth areas – the 'seas' – as well as mountain ranges and craters. The highest mountains are probably the Liebnitz Mountains which are about 26,000 feet in height and more than 30,000 craters have been recorded. Most of these are named after philosophers and scientists such as Archimedes, Copernicus, and Plato. The craters have probably been formed by meteors bombarding the surface early in the Moon's history, perhaps thousands of millions of years ago. Some of these craters are enormous, e.g. Copernicus is 56 miles and Tycho 54 miles across. Their walls are more than two miles in height. On Earth, such meteor craters would be rapidly eroded away by weathering but on the Moon they may last for ever.

The Moon's atmosphere is nearly non-existent and is indeed a more perfect vacuum than can be attained in the laboratory on Earth. This lack of atmosphere on the Moon is due to the fact that its gravity is too low to hold on to any gases. If it ever had any atmosphere it must have lost it at a very early period in its development.

This absence of atmosphere has some interesting effects; shadows are razor sharp; there is no erosion because there is no weather. This means that the Moon's features are more or less as they were at the beginning of time. The footprints left by the first Earth people to walk on the Moon will remain there for all time; they will be there for generations of spacemen, to point the way taken by the early explorers.

The lack of atmosphere also means that the Moon is unprotected from the Sun's rays so that it heats up to 150°C at the equator during the day. At night there is a very sharp fall indeed, again due to lack of atmosphere, and it goes down to −180°C (the temperature of liquid air). Although some micro-organisms could survive they, in common with other living things, could not multiply under such conditions.

Further Reading

K. E. Bullen 'The Interior of the Earth', *Scientific American*, September 1955.

Patrick M. Hurley 'The Confirmation of Continental Drift', *Frontiers in Astronomy*: Readings from *Scientific American*, London 1968.

Marshall Kay 'The Origin of Continents', *Scientific American*, September 1955.

Zdenek Kopal 'The Moon and Man', *Science Journal*, May 1969.

G. P. Kuiper *The Atmosphere of Earth and Planets*, Chicago 1952.

Helmut E. Landsberg 'The Origin of the Atmosphere', *Scientific American*, August 1953.

Larousse Encyclopedia of Astronomy.

R. A. Lyttleton 'The Origin of the Moon', *Science Journal*, May 1969.

Samuel A. Matthews 'This Changing Earth', *National Geographic*, January 1973.

E. J. Öpik 'Climate and the Changing Sun', *Scientific American*, June 1958.

E. J. Öpik 'The Lunar Environment', *Science Journal*, May 1969.

Readers Digest Great World Atlas.

H. C. Urey 'The Origin of the Earth', *Scientific American*, October 1952.

H. C. Urey and G. F. J. Macdonald 'Geophysics of the Moon', *Science Journal*, May 1969.

4. The Beginnings of Life

Where did living things come from?

Early in his evolution Man would have begun to develop a curiosity about his surroundings. At first his speculations would have been concerned with his means of survival – food, shelter, fire, but in time with his increased brain capacity and some increased leisure he would look further afield and begin to wonder about the plants and animals around him. Had they always been? Were they put there by some God or gods? Having no scientific training and no technological tools he found the problems of their origins too much for his brain to cope with and so he turned the matter over to his priests and was content that it should rest there. From time to time through the ages the problem has been aired and discussed, and theories put forward. At first these were purely philosophical but as the scientific method developed so its findings and hypotheses were also thrown into the pool of knowledge.

The possibility that life was created by some superhuman being or God has given rise to great philosophical controversy and has been at the centre of many, if not all, of the religions of the world. This belief is beyond the scope of scientific investigation because it cannot be investigated by scientific methods and tools. You either believe it or you do not. The scientist deals only with phenomena that can be subjected to the scientific experimental method. He conjectures, but then his conjectures lead him on to conduct experiments and to draw deductions from these experiments. Furthermore his experiments must be repeatable by all who care to try them out. The theory that life and Man were created cannot be subjected to experiment and therefore although comforting and philosophically pleasing to many (and possibly correct) is scientifically a dead end.

Related to this belief is the concept of what has been termed the *élan vital* or spirit, the vital 'something' that makes living things what they are and which is capable of an independent existence outside the body. Julian Huxley has said that to attempt to explain living phenomena by a term like *élan vital* is not to explain them at all but merely to give ignorance a name. To state that an amoeba is moved by an '*élan vital*' is equivalent to saying that a locomotive is moved by its *élan locomotif*. Nothing is explained.

The possibility that life has always been on Earth more or less in its present form can be discounted as we shall see by an examination of the fossil record of the rocks. The further we go back in geological time the simpler the organisms become, so that some 600 million years ago we find only a few simple algae and sponges and beyond that there is no record of life at all. It looks as if evolution from the simple to the complex is a real thing. Life has not always been as it is.

Lord Kelvin and many others believed that simple living things came to Earth from other planets and then evolved to higher forms. This idea was supported by the Swedish scientist Svante Arrhenius who worked out a theory on how spores (portions of living matter of simple organisms) whirled by light pressure into the upper layers of the atmosphere of a planet should be able to leave the mother planet and be widely distributed in outer space, ready to be swept up by some other planet.

As we have seen, it is extremely unlikely that there are living things on any of the other planets of the solar system to give rise to these spores. Further we know now that a long time in outer space close to sources of cosmic radiation and ultra-violet light would destroy such spores, no matter how resistant they might be to the intense heat and cold of interstellar space.

Spontaneous generation

The possibility is that living things arose spontaneously from non-living matter. This is a belief that held sway among the more sophisticated peoples of the world for thousands of years, a belief that died with the great Louis Pasteur in the middle of the nineteenth century and has only recently been revived.

It was indeed rather an obvious belief. One could observe with one's own eye that life arose regularly from non-living matter – worms from mud, maggots from decaying meat, mice from refuse of various kinds. This is the view that came to be known as 'spontaneous generation' and it was widely accepted. Virgil for example said that bees were bred in the putrefying entrails of a young bull. Van Helmont (1577–1644) stated, 'The emanations arising from the bottom of the marshes bring forth frogs, leeches, herbs and a good many other things.' He devised a recipe for creating mice, 'Take a dirty shirt and a few grains of corn, place them in a pot and permit them to sit for twenty-four hours. A lively crop of mice will be produced.' It is said that early theologians could comfortably accept this view, for Genesis states not that God created plants and animals but that he bade the earth and waters to bring them forth.

Then spontaneous generation which had held the stage for 2000 years began to crumble. Its edifice was shaken in the seventeenth century by Francesco Redi, an Italian, who performed what now appears to be rather a simple experiment – but all great experiments are simple in hindsight. He placed a piece of net over a tray of freshly killed meat and found that maggots did not develop in such meat but only on the net outside. Maggots did not arise spontaneously from meat said Redi; flies had to lay eggs before they did so.

Experiments such as this and others like them might have killed the theory of spontaneous generation stone dead, were it not for a Dutch draper and maker of fine microscopes – Anthony Van Leeuwenhoek (1632–1723) –

who with his microscopes examined water from ponds and found it to be teeming with living organisms. He opened up a new world, never seen by man – the world of microbes. He founded the science of microbiology, though no comparable microbiologist was to appear on the scene for the next 150 years.

As happens so often, the wrong type of people fastened on to his magnificent findings. Leeuwenhoek did not mention spontaneous generation but its supporters seized on his work – it may be true as Redi had demonstrated, they said, that large animals such as frogs, beetles, and mice did not arise spontaneously from decaying matter – but what of Leeuwenhoek's 'little animals' (microbes), did it not appear that they arose spontaneously? Add a little sugar to water, allow to stand for a period and the water will teem with microbes. This demonstrates, said the believers, that the microbes arose directly from sugar.

Louis Pasteur 'kills' spontaneous generation

Much controversy developed and it was not until 1860 when Louis Pasteur carried out his famous experiments that the spontaneous generation theory for microbes was laid to rest.

He carried out many experiments on this theme – we need describe only one. He boiled broth in a flask plugged with cotton wool and laid it aside to cool. The broth remained sterile for an indefinite period; it did not develop Leeuwenhoek's little animals. Pasteur then removed the cotton wool plug and in a few days the broth was full of little animals. He had demonstrated that the microbes did not develop spontaneously but came in from the air where they existed in enormous (invisible) numbers.

When Pasteur had finished with these and other experiments, nothing remained of the belief in spontaneous generation. Professor Wald has said, 'We tell this story to beginning students of biology as though it represents a triumph of reason over mysticism. In fact it is very nearly the opposite. The reasonable view was to believe in spontaneous generation; the only alternative, to believe in a simple primary act of supernatural creation. For this reason, in the past, many scientists chose to regard the belief in "spontaneous generation" as a "philosophical necessity". Most modern biologists having reviewed with satisfaction the downfall of the spontaneous generation hypothesis, yet unwilling to accept the alternative belief in special creation, are left with nothing.'

The revival of spontaneous generation

Only recently then has science dared to look again at 'spontaneous generation' and to ask 'What did Pasteur demonstrate?' He showed that living organisms did not arise from non-living matter under present day conditions. He did not demonstrate that the origin of living things from non-living matter was impossible under all possible conditions, for instance under conditions that perhaps existed when the world was young.

It is important too at this point to be sure of what we are looking for in the way of living organisms. Not rats and mice and frogs coming into being spontaneously, for these are the present day end-products of long periods of evolution; not even microbes, for these, though small in size, are extremely complex in molecular structure and function. Rather we are looking for organic molecules too small to be seen by even the most powerful light microscopes, molecules that have the power of reproducing themselves bound together in a membrane.

If such molecules were coming into being now, formed by the chance aggregation of simpler substances, we would be unable to see them; further, they would have a short existence for they would be gobbled up by already existing organisms so intense is the struggle for food that goes on in the sea and in the soil.

What are 'living things'?

Before we go on to examine how such 'living molecules' could have come into being let us look at what living things are. In order to do this let us look at some of the discoveries that have been made about living matter.

In the latter years of the sixteenth century Dutch spectacle-makers found that they could adjust two lenses and thus enlarge an object under observation. Such an instrument was called a perspiculum and William Harvey (1578–1657), the famous English physician who discovered the circulation of the blood, used it to watch the heart beats of the wasp.

The important man in our story is however Robert Hooke (1635–1703), an Englishman born on the Isle of Wight. After a spell as assistant to the famous chemist Robert Boyle, Hooke was made curator of the Royal Society in 1663. One of the conditions of his appointment was that he had to produce 'three or four considerable experiments for each of the weekly meetings'. For one of these he brought his improved perspiculum (or microscope as it is now known), the light source of which was a candle.

Among the many things that he examined under his microscope were thin slices of cork. He observed that the cork was not solid but was made up of numerous small compartments like a honeycomb. These he named cells and he calculated that there were more than 1,259,721,000 such cells in a cubic inch of cork! He found that other plant materials were also made up of cells and in 1665 he published his *Micrographia* in which he described these and other findings.

Although his book aroused some interest it was published at a time when people were more concerned with other matters. The Great Plague was ravaging London and the following year the City was to be swept by the Great Fire. Nevertheless the indefatigable Samuel Pepys read and declared it 'the most ingenious book that I have read in my whole life'.

It was later to be realized that what Hooke observed were cell walls – the all important contents were missing because cork is a dead tissue. Nevertheless this was a momentous finding, for others following in his wake were to determine that all living things – seaweeds, mosses, ferns, flowering plants, snails, insects, birds and Man himself are all built up of these tiny compartments which we call cells, and that all living things begin as a single cell – the ovum – which on fertilization divides again and again until the plant or animal or human is completed.

Most cells are small in size, being between one hundredth and one tenth of a millimetre. But some bacterial cells are smaller than this and the largest single cell is probably that of the egg of an ostrich which, when the shell is removed, is about three inches in circumference. It is large because it contains so much reserve food material for the developing embryo. The human egg or ovum on the other hand is only about one tenth of a millimetre because the developing embyro draws its food materials from the mother and not from the egg.

The number of cells of which an organism is composed varies with the species. Some organisms such as the amoeba are composed of a single cell and can be seen only under the microscope. Man on the other hand at maturity is made up of some 60 billion cells, all of them derived from the single celled ovum. In an organism many of the cells become what we term 'differentiated', that is they become modified to perform different functions so that in, e.g. the human, we have nerve cells, muscle cells, liver cells and so on, differing in structure from each other in relation to the work that they have to do.

The cell then is the fundamental piece of all living matter – the building block for all living things and if we are to determine how living things might in the beginning have come into being in this world of ours then the question is: how did this living cell first come into being?

Before going on to examine this question we have to know what we are looking for. What is a living cell?

What are cells?

Cells are made up of a membrane surrounding the essential living matter. In the past this was known as protoplasm, which was giving ignorance a name! We now know a great deal about the structure and function of the various components of the cell and the term protoplasm is tending to fall into disuse.

Proteins

This membrane, made up of a mixture of fats and proteins, is elastic and allows the cell to expand and contract. Proteins are very important in that not only are they the construction materials of the cell but they are also responsible for all the chemical reactions that go on in it.

A living organism has been described as nothing more than a highly integrated chemical factory. This may be an exaggeration but nevertheless the chemical reactions that go on in its cells are highly important; they are under the control of substances which we term enzymes, which are also made up of proteins. These enzymes do not bring about new chemical reactions but they speed up reactions that, without them, would take a very long time. For example, starch left on its own would in the course of time, in hundreds or perhaps in thousands of years, be changed to the sugar glucose. In the presence of the appropriate enzymes it is converted to glucose in the matter of a few minutes. Also glucose would change to carbon dioxide and water, giving off energy in the course of years, but if the correct enzyme is added this reaction will take place in a few minutes. The energy released is then used by the organism, be it Man, monkey or buttercup for its various activities such as growing, reproducing, digesting, and so on.

Proteins are highly specific. This means that a human's proteins are very, very different from a monkey's proteins or from a buttercup's proteins and the more distantly organisms are related, then the more different are their proteins. A human being's proteins are more closely related to the proteins of another human being than they are, say, to those of a monkey. But they are more closely related to those of a monkey than they are to those of a buttercup. Within the human race the more closely related people are then the more their proteins have in common.

This phenomenon lies at the very heart of organ transplant. The actual technique of transplanting a kidney or a lung or a heart is to the competent surgeon a relatively simple matter. The difficulty lies in inducing the patient's body to 'accept' the new organ. Because the proteins of the donor heart differs from the recipient's proteins then the recipient tends to reject it.

This is really a defensive mechanism on the part of the recipient's body. It rejects any 'foreign' proteins and this is the basis of our body's fight against disease organisms such as bacteria. Bacteria are 'foreign' proteins, so that when they enter our bodies then our bodies do their best to reject them. If the rejection process is successful then all is well. If it is unsuccessful then the bacteria gain the upper hand and disease ensues.

In the case of the transplanted organ we want to overcome the rejection process. This can be done by using an organ, such as a kidney, from an identical twin (whose proteins are identical with those of the recipient) or by using special drugs which dampen down the rejection process. The drawback in using the latter is that, although they allow the transplant to 'take', they also allow bacteria and other foreign organisms such as viruses to attack the body with impunity and to cause disease. It is not surprising then that transplant patients on these suppressive drugs are kept under almost sterile conditions.

Proteins are built up of substances known as amino acids. There are about 24 of these known in living things and the number of them, and perhaps more importantly the arrangement of them, in the molecule, determines what type of protein will be produced. Given some 24 amino acids and, being allowed to juggle with them at will, Nature can produce an almost infinite number of proteins. This accounts for the variety of living things on Earth.

Nucleic acids and chromosomes

Another very important group of chemicals to be found in the cell are known as nucleic acids.

Each cell has a globular central body called the nucleus within which we can discern fine threads. These threads are called chromosomes and the number is standard for any one species. Every body cell of a roundworm has 4 chromosomes; every body cell of a fruit fly has 8; of an onion 16; of a crayfish 200; of Man 46. Every reproductive cell has got half the number of chromosomes that are found in the body cells. The ovum of the human has got 23 and the sperm of the human has got 23 so that when fertilization is effected the species number of 46 is restored.

The chromosomes carry the hereditary information from the father and the mother. Characteristics such as number of fingers, colour of eyes, size of feet, sex and so on are all predetermined by the information passed on in the chromosomes. It is not surprising then that attention has been very much focused on these chromosomes and in recent years we have learned a great deal about them.

They are composed of two main elements: proteins which we have already discussed, and nucleic acids. It appears that it is the nucleic acids that carry the hereditary information. Not only do we know their chemical composition but we also know their shape, which is a very important factor.

There are two important groups of nucleic acids, DNA and RNA. DNA is found in the chromosomes and it transmits its messages by a mechanism which we now know in detail to the other nucleic acid, RNA. The RNA then moves out of the nucleus and picking up amino acids it arranges these in an order which is determined by the RNA and forms a protein. Thus the proteins that are formed are determined by the structure of the RNA which in turn is determined by the structure of the DNA in the chromosomes; chromosomes in turn come from the parents.

When an ovum is fertilized by a sperm the resulting cell divides to form two cells. The chromosomes divide at the same time so that the number of chromosomes is kept constant. This is repeated over and over again, always preceded by a doubling of the chromosomes so that their number is kept constant in any one cell and equal in number to the number in the original fertilized ovum.

In time some of the cells begin to differentiate by a process which is not as yet fully understood. In the case of humans some of the cells become liver cells, some heart cells, some cartilage cells and so on, thus forming the appropriate organs in accordance with the 'instructions' coded in the DNA in the chromosome. Similarly, in all living things cells begin to differentiate forming roots, shoots and leaves in plants; wings, legs and bodies in the case of insects, and so on.

Thus two human beings always have human beings as offsprings; mice always have mice; wheat plants always have wheat plants, because the message to form the proteins appropriate to these organisms is passed on in the DNA in the chromosomes.

The above is but a simple outline of what is in effect a very complicated story, which is now well known due largely to the work of two scientists, Crick and Watson, working at Cambridge and for interested readers I would recommend the book *The Double Helix* by J. D. Watson which gives a fascinating, readable account of their discovery.

Energy

If a cell or an organism is to survive it needs energy. It is required to maintain the structure of the living organism, it is required for all the things that living organisms do – eating, walking, talking, reproducing and so on. This energy is got from food but the energy has got to be stored and released as required. The special chemical within which the energy is stored is known as ATP and the energy is released as it is needed from the ATP.

Thus in our search for the beginnings of life we have got to look for proteins (the building blocks and the enzymes of living things), nucleic acids (which carry the hereditary messages) and ATP (the energy-storing and energy-releasing chemical) all bound up together with water, various mineral salts, and a few other organic chemicals, within an envelope or membrane. We have got to look for a living cell.

The primitive world

As we have noted in Chapter 3 the primitive atmosphere of the world was not as we know it today. It was probably made up of ammonia, methane, water vapour, carbon dioxide and nitrogen. Oxygen was absent because of the great avidity for oxygen of many of the metals in the crust of the earth, converting them readily into the corresponding oxides. It has been estimated that even today should oxygen cease to be given off by green plants, all the available free oxygen would be taken up in less than 1000 years by igneous rocks and other incompletely oxidized substances.

Regarding the age of the Earth, we have noted, by measuring the degradation of uranium to lead, that it is probably about 5000 million years old. Early in its existence it was probably a hot Earth, due to radioactivity in the rocks, and water boiling off its surface formed masses of black clouds possibly hundreds of miles thick above its surface. Then as the cooling process took place the rains came, torrential rains that probably continued uninterrupted for centuries, forming the oceans and the river basins. Salts and minerals from rocks accumulated in these oceans. Great lightning flashes rent the skies and the atmosphere was penetrated by ultra-violet rays, x-rays and other radiations from the Sun, from space and from the rocks. This early world was indeed a witch's cauldron and out of the brew were to emerge the first living things.

The theories of Oparin and Haldane

In 1922 the Russian scientist Oparin delivered before the Botanical Society in Moscow a paper which dealt with his conclusions on the origin of living things. He was well aware of Darwin's ideas on evolution and he, in effect, carried Darwin's theory further down the evolutionary line. He proposed that simple inorganic substances such as carbon dioxide, methane, ammonia, water (the primitive atmosphere), gave rise to more complex substances such as sugars, proteins, and fats and these combined in special ways to give rise to the first recognizable living thing. These in turn, as shown by Darwin, evolved into higher living things, including Man.

Oparin viewed the evolution of these groups not as separate phenomena but as steps in the unfolding of one and the same process. Life, he said, is but one of the many rising levels of matter evolving.

Oparin's ideas were first published in 1924 in a booklet titled *The Origin of Life*.

J. B. S. Haldane the British scientist put forward similar ideas in 1928. He suggested that before living things came into being organic compounds 'must have accumulated till the primitive oceans reached the consistency of hot dilute soup'. He based this belief on the assumption that the primitive atmosphere contained carbon dioxide, ammonia, water vapour but no oxygen, and on Oparin's assertion that such a mixture of gases, exposed to ultra-violet light or other radiations, would give rise to a vast variety of organic substances. These radiations acted upon the gases, split them into their constituent atoms which, being washed down by the rain into shallow pools, recombined at random to form more complex molecules. Hence the 'hot dilute soup'.

Central to the life process is carbon. It is a versatile element in that not only can it link up with other atoms such as hydrogen, nitrogen, oxygen, chlorine and so on but equally importantly, carbon atoms can link up with each other to form long chains and rings which can extend into three dimensions, thus forming very complex molecules. Among the compounds that can be formed by these linking processes are various sugars, glycerine, fatty acids, and amino acids.

Of these chemicals possibly the most important from the point of view of the construction of a living organism are the amino acids because as we have seen they are the building blocks of proteins and proteins are the very stuff of life.

Urey, Miller and Fox substantiate the theories

Although Oparin and Haldane put forward their theories in the 1920s it was not until 1953 that their ideas were put to the experimental test. In America Professor Urey and his assistant Miller put a mixture of methane, ammonia, and hydrogen gases with water vapour into a continuously circulating system and subjected the mixture to electrical discharges. The system was allowed to run for one week and then resulting fluid was analysed. It contained a truly amazing number and variety of organic materials, including amino acids. This experiment dramatically demonstrated that a great variety of organic compounds could, as Oparin and Haldane had predicted, be formed from the simple inorganic gases of the primitive atmosphere.

The first major step had been taken. It has been said by Keosian that, 'the metaphysical aura surrounding the problem of the origin of life was thus lifted; serious discussion and experimentations could be and were undertaken.'

Various energy sources such as heat, light, ultra-violet rays, x-rays, radiations, electric sparks and ultrasonic vibrations have subsequently been used by many workers and always similar results have been obtained. The gases have given rise to a variety of organic substances including amino acids.

We know that amino acids are the building blocks of proteins so the question that now arose was: could these amino acids link together to form proteins? Various workers tried heating them together in liquid solutions without success—no proteins were formed. Then Dr Fox and his colleagues in the United States in 1955 tried heating the amino acids together under dry conditions and this did the trick. Primitive proteins were formed. This sort of fusion, said Fox, could have taken place on the hot dry rocks (not in the oceans) of the early world. The proteins could then be washed into the sea. Thus we can get from simple inorganic gases to proteins without man-made (or divine) intervention.

The three major constituents of cells

The development of proteins does not in itself constitute a living organism; proteins cannot reproduce. For this we need the nucleic acids, in particular DNA. These nucleic acids are built up of simple compounds called nucleotides, just as proteins are built of amino acids. It is therefore of interest to note that some of these nucleotides have been synthesized in the laboratory using the types of chemicals that would be reasonably expected to be present in the early world and subjecting them to the types of conditions that prevailed then. Furthermore Gerhard Schramm of the Max Planck Institute in Berlin demonstrated that it was possible to make nucleic acids from these nucleotides by subjecting them to heat. So far however any nucleic acids that have been made have been unable to reproduce themselves. We may perhaps note as an aside that once nucleic acids which can reproduce themselves have been synthesized then the first major step to the creation of life in the laboratory will have been taken.

The third important constituent of a living system – the energy storing and releasing ATP – has also been shown to be synthesized by processes similar to those used in the making of nucleic acids.

It thus appears likely that the three major constituents of living systems as they are known today, protein, nucleic acids and ATP together with various miscellaneous sugars, fats and mineral salts were formed in the primitive world by chance aggregation, perhaps aided by the enzyme action of proteins as they were formed.

Formation of a cell

The next step is the aggregation of these chemicals into an interlocking system enclosed in a membrane. Could this have happened? It seems reasonably likely that it could. Mixtures containing protein-fat complexes can form membranous surface 'skins' similar to the membranes of cells. If one of these skins formed round an aggregate of the other chemicals then the first living cell, the first living unicellular organism came into being. By dividing and separating, this organism would give rise to more unicellular organisms. By dividing and not separating, this organism would give rise to multicellular organisms.

The foregoing account may be the way in which the first living thing arose. It is *not* maintained that this is how it *did* happen (only a fool could be so dogmatic); it is only maintained that this is how it could have happened.

The need for energy – photosynthesis

In order to survive this early organism had to have food, and its early food was probably organic molecules such as traces of sugars, fats and so on that were formed by the same processes by which the organism itself was formed. Such foodstuff would be in very short supply and it would

1. Relative sizes of members of the solar system (Pluto is not shown)

2. Part of the sky seen with the naked eye

3. The same region seen through the telescope

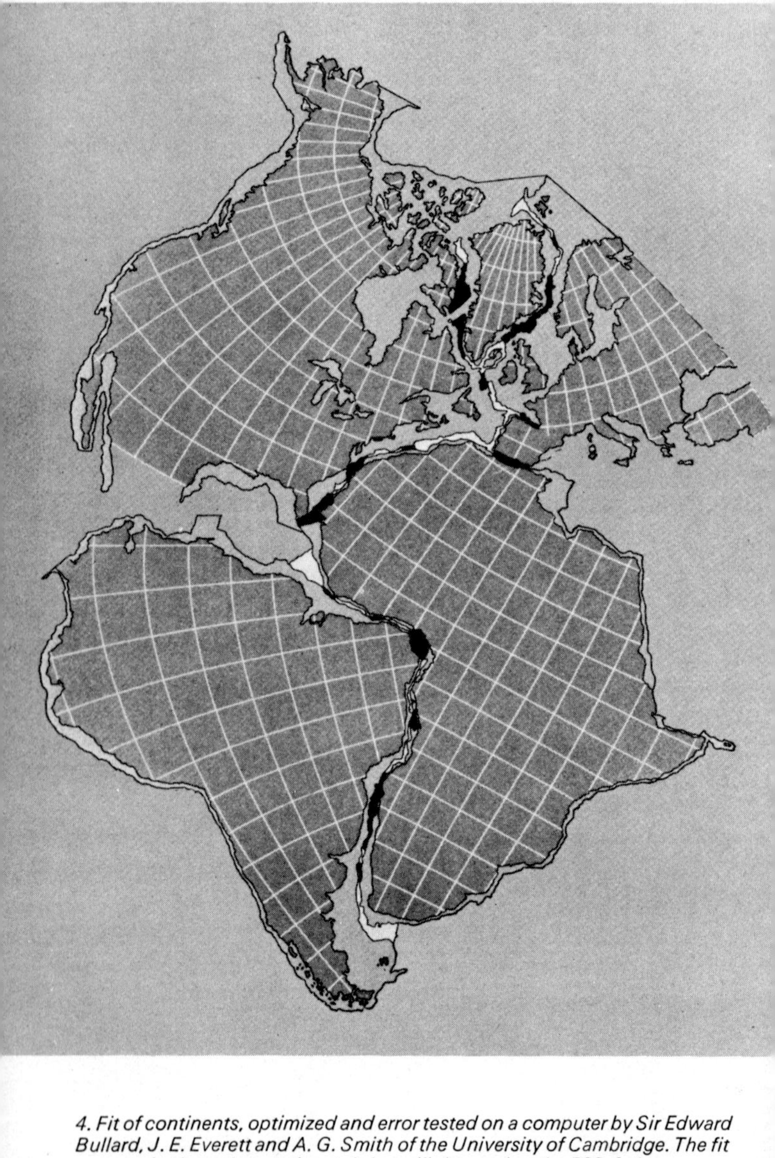

4. *Fit of continents, optimized and error tested on a computer by Sir Edward Bullard, J. E. Everett and A. G. Smith of the University of Cambridge. The fit was made along the continental slope (light grey) at the 500-fathom contour line. The regions where land masses, including the shelf, overlap are black; gaps are white*

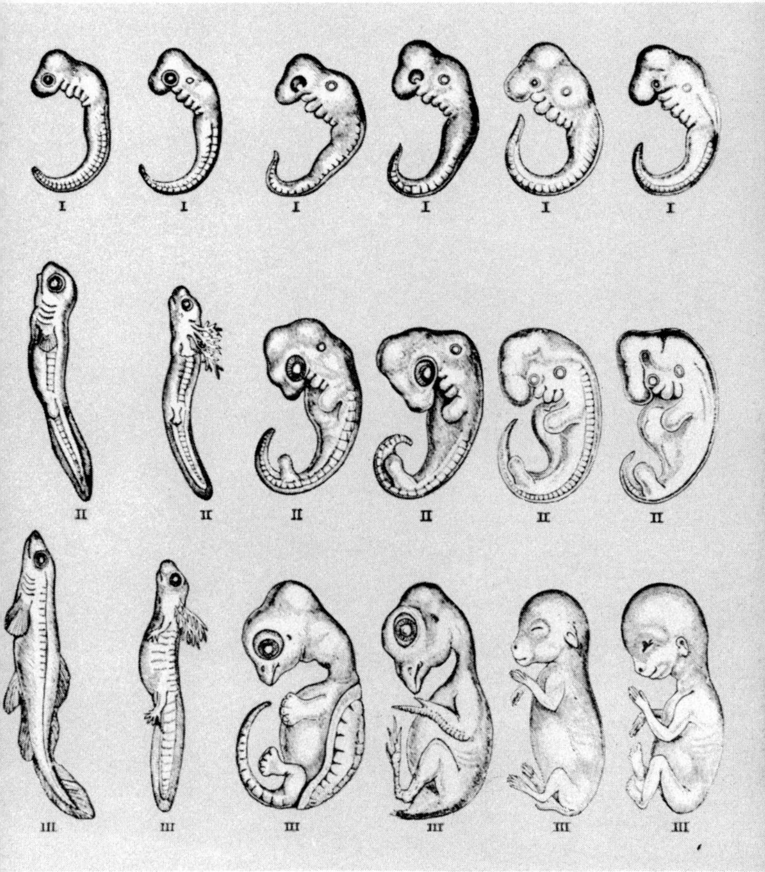

5. Comparison of vertebrate embryos. Three stages in the development of (left to right) a ***fish, salamander, tortoise, chick, rabbit*** *and* ***man.*** *Note close similarity between all early embryos. Note that the more distantly related two vertebrates are, the shorter the period during which they pass through similar embryonic stages*

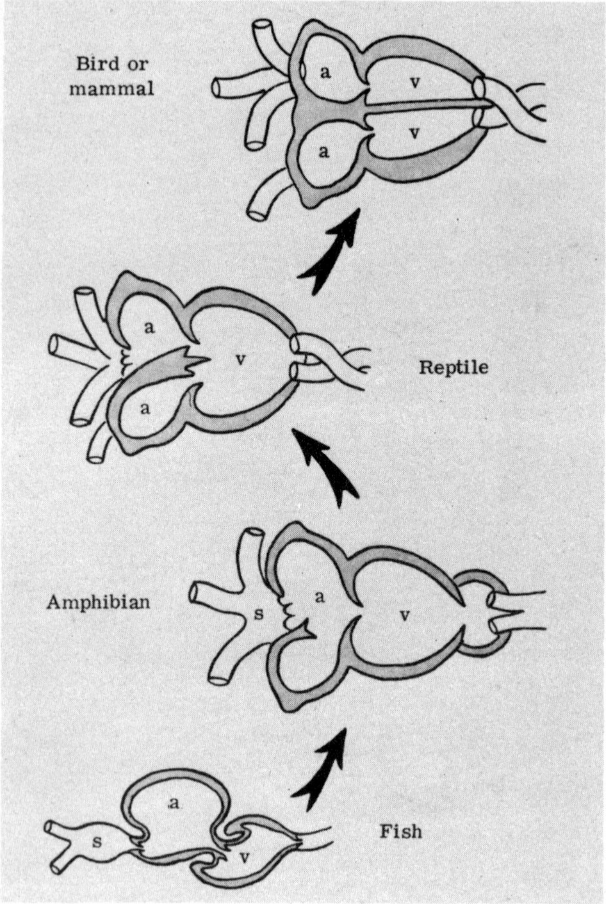

*6. The evolution of the heart. Note one atrium (a) and one ventricle (v) in **fish;** divided atrium in the **amphibian;** divided atrium and partially divided ventricle in **reptile** and divided atrium and divided ventricle in **bird** and **mammal.** These developments reduce the mixing of oxygenated and deoxygenated blood*

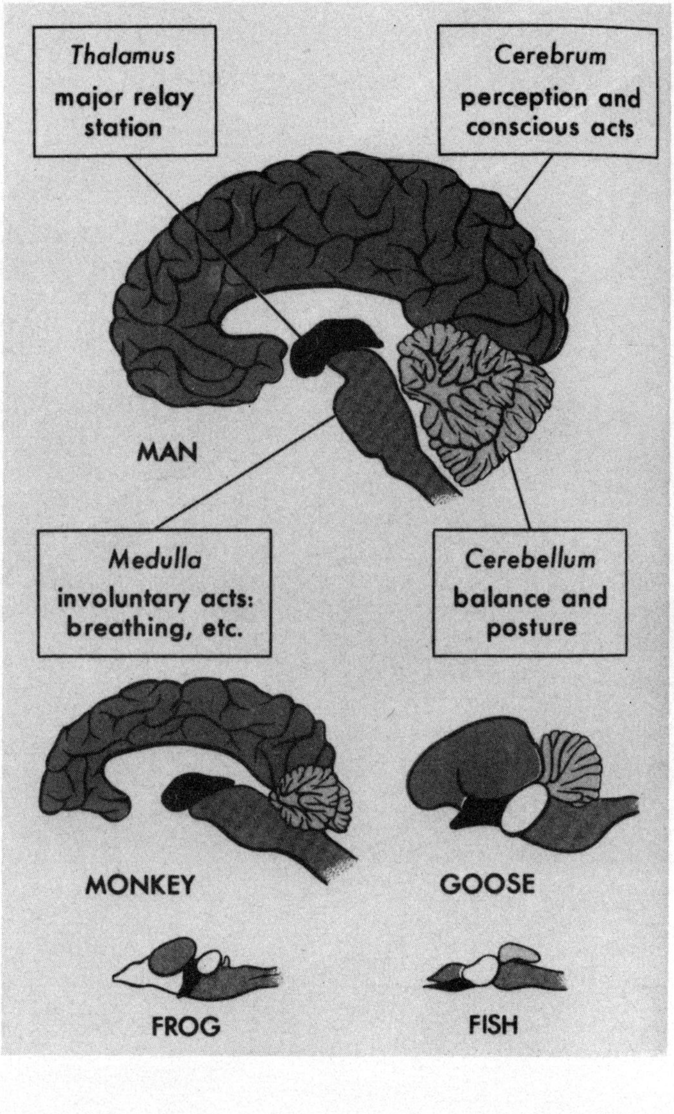

7. A comparison of the major parts of the brain in various animals. Note the greatly developed cerebrum in man and its convolutions in monkey and man

8. The human species: the many varieties of man belong to the same species ***Homo sapiens***

obviously be a tremendous step forward if some of these organisms could manufacture foodstuffs. This they eventually did by developing (with the aid of enzymes) the green colouring matter, chlorophyll.

By the process known as photosynthesis organisms could, in the presence of sunlight, join up carbon dioxide and water to form sugar (as plants do today). The light energy of the sunlight was built as chemical energy into the sugar.

The organisms now had their own food store. However since there was no oxygen, they had in the first instance to gain the energy from the sugar by a process of fermentation similar to that used by yeast today. The process was not very efficient since not all of the potential energy was released by the sugar. Carbon dioxide was released and this was used to build up more sugars but various waste products such as alcohol and lactic acid were also released and these are poisonous to living cells. In the sea many of these products would be washed away but if ever organisms were to move on to land these substances would be toxic since they could not easily be disposed of.

The importance of oxygen

One of the by-products of photosynthesis however is oxygen. Using this, organisms could dispense with the fermentation process and could instead break down the sugar more efficiently by respiration (as most living things do today), thus gaining more energy. The only by-products of this process are carbon dioxide and water. These are not poisonous and are easily disposed of in any environment. Furthermore they could be used over again in photosynthesis to build more foodstuffs. Thus the organisms could get as much energy as they required.

The entry of oxygen into the atmosphere liberated organisms in another way. The sun's radiations contain ultra-violet components which kill off living cells. Water is an effective filter for these radiations so that early life developed in the water because planet Earth was subjected to these ultra-violet radiations. If ever organisms were to move on to land however they would have to be protected. Oxygen was responsible for this protection, which is with us to the present day. Some of it, rising high in the atmosphere, was converted by the radiations to ozone and this ozone layer formed an effective barrier to the radiations! Now organisms could emerge from the water and populate the Earth. It is sometimes said that it is remarkable how well adapted organisms are to their environment. One might remark that they ought to be as in many cases they formed it.

Without photosynthesis life might have continued at best within a very narrow cycle. Even with photosynthesis organisms could lead only a marginal existence with fermentation. The advent of oxygen made respiration possible, which combined with photosynthesis made all things possible. Wald has said, 'it raised organisms above the subsistence level; it provided them with capital which they invested in the great enterprise of organic evolution.' A cyclic process was set in motion of an efficiency which was to set the great evolutionary process in train, resulting in the myriad of organisms that inhabit our lands and waters today and which as we shall see was to culminate in the development of Man himself.

Further Reading

J. D. Bernal *The Origin of Life*, The Weidenfeld & Nicolson Natural History Series, 1967.

J. D. Bernal and Ann Synge *The Origin of Life*, Oxford Biology Readers, J. J. Head and O. E. Lowenstein (Eds), 1972.

G. Ehrensvard *Life ; Origin and Development, Chicago*, 1962.

G. Hardin *Biology: Its Principles and Implications*, London 1961.

A. I. Oparin *The Origin of Life on Earth*, Edinburgh 1957.

G. Wald 'The Origin of Life', *New Treasury of Science* pp 416–428.

5. Evolution in Action

The voyage of the Beagle

In the autumn of 1831 two young men dined together in London. One of them was Robert Fitzroy aged 26, the Captain of HMS *Beagle*, a ship of some 235 tons. Fitzroy was planning a trip round the world to map hitherto uncharted areas and to gather scientific information about these areas. He was attempting to persuade the other young man, Charles Darwin, to come with him as ship's naturalist. Darwin was only 22, a 'gentleman idler'. Although he had graduated in classical studies from Cambridge he had already failed in medicine at the University of Edinburgh and his family were hoping, almost in desperation, that he might become a country parson. He was the son of a wealthy physician of Shrewsbury who apparently did not see any signs of future greatness in his son. 'You care for nothing but shooting, dogs, and rat-catching and will be a disgrace to yourself and all your family.'

Darwin was undecided about the proposed trip. It would take five years and five years was a long time to be away from the centre of things. Finally he was persuaded, though he obviously had misgivings. He wrote to his sister thus, 'Fitzroy says that the stormy sea is exaggerated; that if I do not choose to remain with them, I can at any time get home to England; and that if I like, I shall be left in some healthy, safe and nice country; that I shall always have assistance; that he has many books, all instruments, guns, at my service. There is indeed a tide in the affairs of men, and I have experienced it. Dearest Susan, Goodbye.'

And so HMS *Beagle*, rigged as a barque and carrying six guns, set sail from Devonport harbour on 27 December

1831. The voyage almost ended before it had properly begun. Fitzroy had not been correct in telling Darwin that the stormy seas were exaggerated! On the contrary! *Beagle* ran into high seas which almost swamped her and Darwin suffered agonies of seasickness.

In due course however the ship reached South America and he had the opportunity of studying the native plants and animals there. From South America the ship made her way to the Galapagos Islands, lying directly on the equator, some 600 miles off the west coast of South America. Galapagos had at one time been a refuge for buccaneers and the islands were little more than the 'chimneys of burnt-out volcanoes'. Darwin was however astonished at the variety of animal life to be seen here – giant armoured tortoises that trundled their way through the undergrowth, sea lizards, more than three feet in length, that lay sunning themselves on the beaches. 'There is no other quarter of the world where this order (reptiles) replaces the herbivorous mammalia in so extraordinary a manner.'

He was especially taken with the little finches – they occurred in such a variety of forms, in particular different varieties had different shapes of beaks. Finches normally are seed eaters and Darwin noted that the ground finches on Galapagos had broad heavy beaks for cracking seeds. However the tree finches had short thick beaks and ate insects. One species of tree finch had a longer, straighter beak like a woodpecker's but it had no long tongue like a woodpecker and it had evolved the habit of using a cactus spine or a piece of wood as a substitute to probe out the insects from the cracks in the bark of the tree. Others had parrot-like beaks and fed on fruit. Another strange thing he noted. The finches were practically the only land birds there. It was almost as if they had been produced in such variety to take the places of other species of birds that could be found on the greater land masses. He also noted that on separate islands the same species of animal showed differences in structure and colour. In all living things he saw this great variation and as we shall see variation was to be the cornerstone of his theory of evolution.

The importance of variation and selection

In October 1836, the *Beagle* having sailed round the world returned to England. Darwin was 27 years of age and he was already feeling the beginnings of the ill-health that was to plague his days. He brought back with him a mass of notes and data on the various things that he had seen on his voyage.

Always at the front of his mind was this grand spectacle of variation that he had noticed over and over again on his voyage and now back home he proceeded to collect evidence of variation in London from the breeders of new varieties of pigeons, fowls, dogs, cattle and horses. He himself experimented in raising and breeding domestic varieties of pigeons and he found that by mating pigeons with different characteristics he could produce completely new varieties of pigeons. From his own work with pigeons and from the information supplied to him from breeders he noted that for the production of new varieties of domestic animals two factors were necessary. First, the animal must produce variations and, secondly, the breeder must select from among the many varieties those that he wishes to breed true.

Once more Darwin turned his thoughts back to plants and animals in nature. There was no doubt about the variation in any one species; he had noted these in his travels. There was however no breeder in Nature; where and what was the selector?

At this time he happened upon an essay written in 1798 by the Reverend T. R. Malthus, an English minister and sociologist, who proposed that human beings reproduce themselves many times more rapidly than they can increase their necessary food supply. As a result there is a struggle for existence among members of the human race. In his autobiography Darwin states, 'In October 1838, that is, fifteen months after I had begun my systematic enquiry I happened to read, for amusement, Malthus on "Population" and being well prepared to appreciate the struggle for existence which everywhere goes on from long-continued observation of the habits of animals and plants, it at once struck me that under these circumstances favourable variations would tend to be preserved and unfavourable ones to be destroyed. The result of this would be the formation of new species. Here, then, I had at last got a theory by which to work; but I was so anxious to avoid prejudice that I determined not for some time to write even the briefest sketch of it.' Darwin had realized that the selector was Nature herself. His deduction was that species, contrary to almost universal belief, are not fixed and unchanging but actually do change over long periods of time. These thoughts were running through his head in London in 1838. He was then 28 years of age.

A period of thinking

In January 1839 Darwin married his cousin Emma Wedgewood, a grand-daughter of the founder of the well-known pottery firm of that name, and in September 1842 they moved their household out of London to a little village, Downe, in West Kent. Having inherited money, Darwin was able to devote himself to the question that was never far from his mind – evolution. His voyage round the world seems to have satisfied any desire that he had had for travel. From Downe he never went further than the neighbouring watering places for the sake of his health.

In June 1842, before leaving London, he had filled 35 pages of manuscript with an outline of his *Origin of Species*. By 1844 he had drawn much of his evidence together and had filled 230 pages of manuscript. But still he had not published anything. He continued writing from time to time but the next seven years were taken up mainly with his research on barnacles.

The years passed and he preferred to talk about his

work with his chosen circle of friends. One of them Lyell, the great geologist, who had much influenced Darwin's thinking, warned him in 1856, 'You will be anticipated, you had better publish.' Of Darwin, his gardener said, 'Poor man, he just stands and stares at a yellow flower for minutes at a time. He would be better off with something to do.' And thus the gardener joins with Darwin's father in being responsible for two of the most mistaken judgements in all history.

Many theories have been put forward as to why he did not publish during this long period. Some have felt that it was out of deference to Captain Fitzroy who was a very religious man and who might have been deeply hurt by the outcome of his naturalist's findings. Others put forward the idea that Darwin, anticipating that a storm would break about his head when the implications of his theory were realized, preferred to delay as long as possible. He indeed left instructions to his wife that she was to publish his work if he died.

A letter from A. R. Wallace

And then like a bolt from the blue Lyell's prediction came true. Alfred Russell Wallace some fourteen years younger than Darwin, a naturalist and collector, had visited South America and while collecting in the Amazon had been struck by the great variability of individuals within a species. On arrival in Borneo he had read Malthus' *Essay on Population* as Darwin had done eighteen years before. It is said that whilst suffering from a bout of malaria his fevered brain had hit upon natural selection of variable organisms as the driving force of evolution. He penned a letter to Charles Darwin in the summer of 1858 and his sketch was almost a replica of Darwin's own outline written in 1842.

Darwin on receiving this letter must have felt that the bottom had fallen out of his world. By now he was 50 years of age and 22 years of work had been for nought, and his first impulse was to withdraw totally in favour of Wallace. 'I would far rather burn my whole book than that he or any other man should think that I had behaved in a paltry spirit.' He was dissuaded from doing so by his great friends Lyell and Sir Joseph Hooker, the botanist. They asked him to let them have an outline of his theory and this together with Wallace's outline they sent to the Linnean Society of London.

The two papers were read at a meeting before the Fellows on 1 July 1858. It would appear that although the communications raised some interest they did not cause a stir. Darwin went on to finish his book and the *Origin of Species* was published on 24 November 1859. His publishers, like the Fellows of the Linnean before them, did not appear to realize the tremendous implications of this book. Only 1250 copies were printed of this first edition. It was sold out in a single day! More editions followed in 1860, 1861, 1866, 1869 and 1872.

People came to realize that they had in their hands the most revolutionary book since the Bible and that the world of Nature would never look quite the same again. Thomas Huxley, the famous biologist, who was to be one of Darwin's great champions in the, at times, stormy days ahead said after reading it, 'How extremely stupid not to have thought of that.' But aren't all discoveries simple after some genius has discovered them?

The central themes in Darwin's theory

It is worth emphasizing the central themes in Darwin's theory. They are that organisms, plant and animal, increase in number at a geometric rate from generation to generation. Geometric ratio can be illustrated by the amoeba. If 1 amoeba divides to form 2, then those 2 will divide to form 4; 4 will divide to form 8, then 16, 32, 64, 128 and so on. A geometric ratio is produced by multiplying, not by adding.

Darwin reasoned that each generation of any organism has more offspring than parents. Despite this however there is no apparent marked increase in the numbers of individuals in any given species. He couldn't see any gigantic rise in population numbers. The number remains fairly constant. There must therefore be a struggle for survival, there must be something controlling population numbers. And this 'something' he realized is competition for food, for water, for light and so on. Many organisms in any one species do not survive this competition. Only the 'fittest', i.e. those most suited to the environment, survive. Among individuals in any one species variations are found and many of the variations are inherited. Some are favourable to an organism in a particular environment and help it to survive so that it may reproduce. Surviving organisms pass on these variations to their offspring until, with the accumulation of a great many variations, such individuals are so different from the originals that they can be considered a new species.

The history of evolution – Lamarckism

Darwin's book was the first demonstration that the law of evolution holds good for every living thing. He was not however the first to proclaim that evolution takes place. The Greek philosopher Anaximander of the fifth century BC believed that men were first formed as fishes and then cast off their fish skins and took up life on dry land. Xenophanes, also of the fifth century BC, is credited with being the first person to recognize that fossils represent the remains of animals that once lived. The fourth century BC saw the 'father of the evolution idea', Empedocles, who taught that plants and animals and Man rose out of the earth as unattached organs and parts which joined together in haphazard fashion. Most of them were freaks and monsters incapable of living, but occasionally a combination appeared which survived and populated the earth. Is this the first glimmerings of variation, natural selection and survival of the fittest? The fourth century BC saw the rise of Aristotle who carried out investigations in various fields of biology. He maintained that there is a complete gradation in nature. The lowest stage is the inorganic. Organic beings arose from inorganic by direct metamorphosis. Within the animal group he constructed a series from the lowest form leading right up to Man. His 'tree' had no branches – it was a straight line to Man.

In more modern times Darwin's grandfather, Erasmus Darwin, in 1794 published a long paper, *Zoonomia*, in which he stated his belief that evolution occurred in plants and animals but he could not propose any mechanisms for it. A French biologist Jean-Baptiste Lamarck could and did in his book *Zoological Philosophy* published in 1809, the year Darwin was born. He proposed that species change as a result of a *need* for change. The need was brought about by the animals' 'inner feelings'. He proposed the law of use and disuse – the more a part of the body is used the more it develops; the less it is used the more it weakens and tends to disappear. He also put forward the 'law of inheritance of acquired characteristics' assuming that an animal could pass on to its offspring characteristics that it had acquired during its lifetime. For example he reasoned that the ancestors of snakes had legs like lizards but they used these legs less and less because of crawling and over a long period the legs disappeared. Some birds developed webbed feet because they swam so much; giraffes got long necks because they stretched so much for food and these characteristics were passed on. It all seemed very reasonable but repeated experiments have failed to show that acquired characteristics are passed on from generation to generation. For example, the tails were cut off twenty generations of mice. The twenty-first generation still had tails. No matter how deeply tanned with the sun people may get, their children will still be born white.

It is important to differentiate between Lamarck's theory and Darwin's. Lamarck said that snakes lost their legs because they stopped using them. Darwin would say that some reptiles lost their legs through a chance mutation which produced variants that had no legs and that adopting a new mode of life found this loss to be advantageous. The new species that thus evolved, snakes, therefore survived.

Where however Darwin differed substantially from all the other proponents of evolutionary theories was that he backed up his theory with masses of facts. He let the facts speak for themselves. Wallace writing in 1887, five years after Darwin's death, said, 'I had the idea of working it out so far as I was able when I returned home, not at all expecting that Darwin had anticipated me. I can truly say now, as I said many years ago, that I am glad it was so for I have not the love of work, experiment and detail that was so pre-eminent in Darwin and without which anything I could have written would never have convinced the world.' Most important of all Darwin put forward a theory to explain how evolution occurs. This he called 'natural selection'.

The birth of genetics

In the years that have followed as we shall see as more and more facts have been uncovered these have strengthened Darwinism (as Wallace called it). And yet in Darwin's time there were two questions that he could not answer. One was, 'What is the source of these variations that occur?'; and the other, 'Can you show us any examples of natural selection actually in action?'

The answers to these questions had to wait for later generations and the scene moves rather improbably to Czechoslovakia, to a monastery garden looked after by a monk, Gregor Mendel. Mendel was a monk in the Augustinian monastery at Brünn, Austria (now Brno, Czechoslovakia). He came as a poor boy to the monastery in 1843 and was ordained a priest in 1847. In 1851 he was sent by his order to study natural science at the University in Vienna. He did not do particularly well and returned to the monastery as a science teacher in 1854. In his spare time he began in 1857 to do breeding work with peas in which he crossed varieties with contrasting characters. For example he crossed tall pea plants with short pea plants and noted that all of the offspring were tall; he then crossed these tall offspring with each other and noted that some of the offspring were tall and some were short. Furthermore he counted the numbers of each. With these and similar experiments using other contrasting characters he developed laws of breeding by means of which he was able to predict not only the characteristics of offspring but also the proportions in which they would occur. In so doing he laid the foundations of modern genetics.

His findings also led him to put forward the theory that every characteristic in the pea plant was determined by a pair of 'factors' or, as we now call them, 'genes' so that each pea plant has a great many characteristics because it has a great many sets of genes. Two genes determined flower colour; two genes determined the size of seed, and so on. One of each pair of genes, said Mendel, is got from one parent and one from the other.

For more than 35 years Mendel's findings were neglected and then in 1900 they were discovered independently by scientists in Germany, the Netherlands and Austria. His work had to wait until the scientific world was ready to receive it. Since that time genetics – the study of heredity – has been one of the most active branches of biological science. Mendel's Laws were found to hold true (under certain circumstances) not only for peas but for all living things including Man. We now have a deep knowledge of the genetics of a great many organisms. The full potential of the knowledge gained has yet to be applied but already the laws of genetics are used extensively in plant and animal breeding.

It was however a young graduate student at Columbia University in the USA who suggested in 1902 that Mendel's 'factors' or 'genes' were actually located in the nucleus of the reproductive cells; that the genes were linked together to form the thin threads that could be seen under the microscope, the threads which we call chromosomes. The chromosomes indeed were composed of genes as a pearl necklace is composed of pearls.

Numerous experiments since then have shown that the hereditary characteristics of all living things, not only peas but all plants, not only flies but all insects, not only animals but also Man, are carried and transmitted from generation to generation by means of the chromosomes. The number is constant for any one species; thus every body cell of every human being has got 46 chromosomes, except for the reproductive cells which have got only half the number. When two reproductive cells join to produce a new being the number 46 is thus reconstituted. It has now been firmly established that the hereditary characteristics of all living organisms are determined by genes and these genes are carried on the chromosomes.

Mutations

And what has all of this to do with Darwin and evolution?

We now know that these genes and chromosomes sometimes change their structure and as a result of this the characteristics of the organism may alter. These changes can be produced by means of x-rays and other types of radiations, by chemicals, by heat and by other factors. Thus a red-eyed fly subjected to x-rays may have a number of offspring with white eyes and these white-eyed offspring will breed true, i.e. they will produce white-eyed offspring. These changes are known as 'mutations'. Sometimes the changes come about without any obvious factor being at work. These we term 'spontaneous mutations'. Those brought about by some agency employed by Man we term 'induced mutations'. The offspring of both are known as 'mutants' and mutants are the very stuff of evolution.

Most mutations are harmful because most organisms are in tune with the environment in which they live but should the environment change or should new environments be available then the mutations may be advantageous and thus incorporated in the race. For example if a mutation in certain fish produced lungs instead of gills then this mutation would be harmful (i.e. the fish would die) if the fish lived deep in the Atlantic Ocean. If however the fish lived close to land then the same mutation might be very beneficial since it would enable the fish to invade the new dry environment.

A nice example of the importance of the environment is given by Karl von Frisch in his discussion of Darwin's theory in the light of modern genetics (see Further Reading). Almost all insects at some time or another give rise to mutant offspring which are wingless. Under normal conditions these insects are at a serious disadvantage with their winged counterparts; they perish because they cannot survive in the struggle for existence. The mutation is harmful.

On an isolated wind-swept island however the position may be reversed. Such an island is Kerguelen in the Indian Ocean where all the native insects including many species of flies, butterflies and moths are all wingless and incapable of flight. Presumably they also arose as mutants from normal winged ancestors but instead of finding winglessness a disadvantage in the struggle for survival they find it an advantage. Any insects with wings rise in the air and are all too readily carried out to sea and die. Thus the winged forms perish while the wingless forms survive and breed and in time these wingless varieties will be considered new species.

One cannot think of a better example of the survival of the fittest, of variants being chosen for evolutionary advance by natural selection. So now, even if Darwin didn't, we know the source of his variations – they are mutations in the genes – and they further strengthen and support his theory. We are now able to describe many examples of natural selection in action. One we have given above; another concerns the celebrated case of the peppered moth.

In the latter half of the eighteenth century the industrial revolution got under way in England and with it came what is sometimes regarded as a modern phenomenon, pollution of the atmosphere as the chimneys of the 'dark satanic mills' poured out their belching smoke. In some industrial areas the resulting 'fall out' of smoke particles was enormous, amounting in some cases to 50 tons per square mile per month. As a result the trunks of trees in the neighbourhood were blackened and sooty in colour. And then an interesting thing was noted; many of the moths, formerly light coloured, using these sooty trunks as resting places were also dark, and this dark colouration was passed on to their offspring. Was this Lamarckism in action? Was he right after all? Could acquired characteristics really be passed on? Similar changes in the moth population were noted in the industrial areas of France. Germany, Poland, Czechoslovakia, Canada and the United States of America. About 90% of any one species were dark, only 10% remained light. The condition became known as 'industrial melanism' and it was noted only on those moths that flew at night and rested by day on the dark tree trunks.

Investigation work was undertaken at Oxford University with one of these species, the peppered moth. Equal numbers of dark and light forms (both specially marked so that they could later be identified) were released into a polluted woodland. Collections were made some days later when it was found that nearly all of the surviving collected moths were dark coloured. Similar mixed numbers were released in an unpolluted wood where the tree trunks were light coloured. In this case nearly all of the moths collected later were light coloured. This was puzzling. Then it was noticed that birds fed on the moths by picking them off the trees. The light coloured moths stood out clearly against the darkened tree, were eaten, and so the dark coloured moths survived. On the unpolluted light coloured trees on the other hand the dark moths were readily visible and were easily picked off by the birds and the light coloured moths survived. It was not Lamarckism at all. It was Darwinism in action. It was an example of Nature in the shape of the birds acting as a selector on variants. It was natural selection in action.

The evidence in support of the theory of evolution

There is now an overwhelming mass of evidence in support of evolution. It comes from many quarters and from many disciplines. Some of the earliest evidence came from the rocks and it continues to accumulate as more and more fossils are found. Fossils are the remains of a once living organism. It is rare to find the organism wholly preserved but there are some such insects which were trapped in the gum from certain trees growing along the Baltic coast more than 30 million years ago. The gum solidified quickly to form a hard glassy-like amber and the insects trapped inside are as well preserved as when they were alive. Ice can also be an excellent preservative. Frozen mammoths which must have died more than 20,000 years ago were found in Siberia in the early years of this century. The meat was well enough preserved to be fed to dogs.

Such total preservation is however rare and more usually we find only the hard parts of animals such as bones and shells. Some of these may be more than 500 million years old and they yield valuable evidence of what life was like long ago. In some cases a perfect mould may be formed due to the organism pressing on the soft mud. Later as the organism decays the mould may fill up with mineral matter and thus we get a very good cast. Many plant remains have been well preserved in this way and we have also many imprints left by dinosaurs more than 120 million years ago.

It is only rarely that fossils are the same as creatures now living on Earth and after much examination, searching and debate it has been concluded that they represent plant and animal remains in an earlier state of evolution. The further that we go back in the rocks the simpler the fossils become.

It should be realized that only a very few organisms ever

form fossils. The conditions must be favourable for preservation or the organism deteriorates without leaving a trace. Furthermore most of the fossils that have ever been formed undoubtedly lie locked away under the great mountain ranges, perhaps never to be revealed. It is not surprising that we have but an incomplete record from the rocks.

Further evidence in support of evolution comes from an examination of the organs of various animals. We find that the forelimbs of human beings, of the whale and of the bat are all very similar although they have very different functions, namely lifting, swimming and flying respectively. This points to these organs having a common origin because it seems hardly likely that the same structure would best meet the needs of all these different functions. We note that both the boa constrictor and the whale have got hip bones which have no function in locomotion. They are organs left behind in the evolutionary process.

Some of the strongest evidence comes from embryology – where we find that the embryos of all vertebrate animals (i.e. those with a backbone) such as the fish, salamander, tortoise, hen, rabbit, and Man are almost indistinguishable from each other at an early stage of development. Furthermore at this early stage all of them, including humans, have got primitive gill clefts or slits.

It is now recognized that in their embryological development all animals (and plants too) go through some of the early stages of their evolutionary history because they have all inherited their developmental mechanisms from a common ancestor. The more closely related any two vertebrates are then the longer their embryonic development proceeds in a parallel fashion.

It is not only in structural matters that we find this resemblance. It is also shown in their chemical make-up. Fish can get rid of waste nitrogenous materials as ammonia, while frogs can excrete them as urea. We find that the fish-like tadpole stage of the frog excretes the waste as ammonia until it metamorphoses into the adult frog, when the excretion changes to urea. There are many other examples of biochemical similarities e.g. all organisms, plant and animal, carry and transmit their hereditary factors through nucleic acids. The amoeba and Man and the sunflower have got this common mechanism of inheritance.

Further evidence is got from the geographical distribution of plants and animals. No frogs are found on the Hawaian Islands, no snakes in Ireland. Rabbits were not found in Australia till they were introduced there by Man. The only reasonable explanation for this discontinuous distribution would seem to be that different types evolved in different places and their distribution was limited or inhibited by geographical factors such as water or mountains. Geographical separation whether by mountains, rivers or oceans can be an important factor in the development of species. It gives the new mutants an opportunity to breed and develop without being 'diluted'. Thus we may note that the mammals found in Australia are unique to that part of the world, indicating a separation of that continent from the mainland at an early stage in the evolution of mammals. Likewise the birds and mammals of Africa are different from those of South America. We find tigers as representatives of the cat family in India and Asia; pumas in South America. Likewise ostriches are found in Africa, rheas in South America and emus in Australia.

Finally man's own activities, as Darwin noted, in the breeding of domestic plants and animals are an example of evolution in action. Over the last few thousand years and particularly in the last 200, by breeding and selection, he has produced thousands of varieties of food plants and animals, as well as domestic pets such as the dog, ranging from the Irish wolfhound to the chihuahua.

Further Reading

James F. Crow 'Ionizing Radiation and Evolution', *Scientific American*, September 1959.

E. S. Deevey 'Living Records of the Ice Age', *Scientific American*, May 1949.

Theodosius Dobzhansky 'The Genetic Basis of Evolution', *Scientific American*, January 1950.

Loren C. Eisely 'Charles Darwin', *Scientific American*, February 1956.

Karl von Frisch 'Darwin's Theory in the Light of Modern Genetics', *New Treasury of Science*, p 467.

Julian Huxley *Evolution, The Modern Synthesis*, London 1955.

Catherine Jarman *Evolution of Life*, London 1970.

Sir Arthur Keith 'Darwin and *The Origin of Species*', *New Treasury of Science*, p 447.

H. B. D. Kettlewell 'Darwin's Missing Evidence', *Scientific American*, March 1959.

John W. Kimball *Biology*, 2nd Edition, Reading (Mass) 1968.

P. A. Moody *Introduction to Evolution*, 2nd Edition, London 1964.

Alan Moorehead *Darwin and the Beagle*, London 1971.

Molecules to Man. Biological Sciences Curriculum Study, Blue Version. Boston 1968.

6. The Procession of Life — Animals without backbones

In the next three chapters we shall follow the process of evolution from simple marine animals right through to Man himself. Some of the material that is presented is factual, there being actual living or fossil links; some is conjectural because we have not found the missing evidence; but it is hoped that enough will be written to convince even the most sceptical that this amazing, inspiring story of evolution is a true one.

As more and more evidence comes in from a variety of sources ranging from the biochemists' laboratories to the graves of prehistoric Africa so the pieces fit together to reveal this awe-inspiring sight – the procession of life.

We shall note that as we move up the evolutionary scale myriads of changes take place – some small, some large, all important.

Out of the vast array of changes and developments, two are particularly worthy of note. These are the development of the heart to meet the increasing need for energy flow and the development of the complex nervous system, including the brain, to meet the challenges of the environment. It is perhaps worth spending a few minutes looking at these.

The heart

Essentially the heart is a pumping organ to send the blood speeding round the body. In Man it is a fairly complex organ but it had simple beginnings. In worms all the blood vessels pulsate regularly driving the blood along. In molluscs the heart is a single chamber called the ventricle. This heart pumps blood into vessels whose branches end in open tissue spaces. There are valves connected with the ventricle to prevent backflow. The circulation is sluggish and some molluscs have developed a series of booster pumps in the form of small accessory hearts to help the blood along.

Fish have developed a second chamber, the atrium, in addition to the ventricle. Blood passes from the atrium to the ventricle to the gills, to the body tissues and is returned to the atrium through a series of veins. We find that the amphibians such as the frog have taken another step forward in that the atrium is divided into two chambers receiving de-oxygenated blood from the body into the right atrium and oxygenated blood from the lungs in the left atrium. Although the two bloods are then passed into a single ventricle for onward passage to the body there is not a great mixing of the two.

To get a really efficient system it would be desirable to have the ventricle divided into two as well so that blood coming from the lungs could be pumped through the body and blood coming from the body could be pumped to the lungs for oxygenating. We find the beginnings of this division in the reptiles, particularly the crocodile, but it reaches complete development in the birds and the mammals which have all got four-chambered hearts.

The brain

Man's brain – 'the great ravelled knot' – weighs about 50 ounces (male) and 45 ounces (female) and consists of thousands of millions of cells; its complexity of structure is almost unbelievable. Nevertheless it too had simple beginnings, starting as a few simple nerve cells in the sea anemones. Development has proceeded further in the jellyfish and we find that at certain places some of the nerve cells have grouped together to form ganglia. Moving on to the worms we find that the more lowly developed forms have got two nerve cords with a pair of swellings at the ends to form a primitive brain. The squid and the octopus have got well developed brains which form a ring round the oesophagus – compact, complex and by far the largest brain to be found among the animals without backbones.

It is however when we come to the backboned animals that we find the major developments taking place. The brain can be regarded as the forward end of the spinal cord which has become expanded to meet certain needs. The brain of the dogfish for example is made up of the following four parts (proceeding from front to rear).

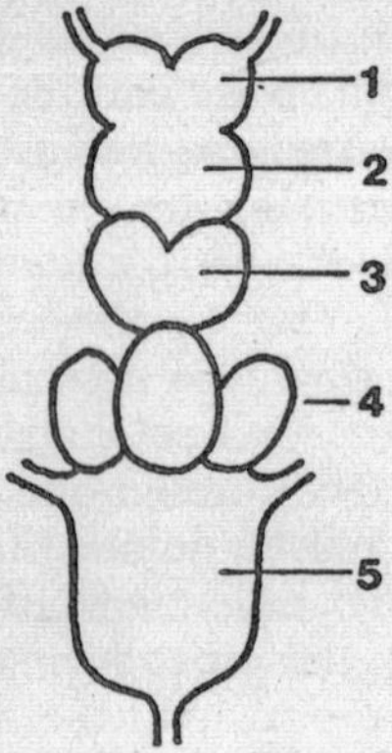

1. a pair of olfactory lobes dealing with sense of smell
2. cerebral hemispheres – inconspicuous
3. a pair of optic lobes, each connected to an eye
4. a region called the cerebellum associated with balance
5. a posterior region, the medulla, which is clearly an expansion of the spinal cord.

It may be noted that the cerebral hemispheres are small and inconspicuous. These are co-ordinators or inhibitors, which as we shall see become the largest and dominant part of the brain of terrestrial vertebrates. The brains of amphibians and reptiles show enlargement of the cerebral hemispheres and they are relatively much larger than the optic lobes. The brain of both birds and mammals is enormously enlarged compared with the reptiles. This is due mainly to the development of the cerebral hemispheres.

In the higher vertebrates, the cerebral hemispheres in front and the cerebellum behind become greatly expanded. The relatively insignificant cerebral hemispheres of fish have given place to a large and elaborate structure made up of three components (1) the cortex or skin of grey matter (2) white matter below it, consisting of nerve fibres connecting the cortex with lower brain centres and the spinal cord and (3) the tissue which links up the left and right sides of the brain.

In Man, when the cerebral cortex is fully grown, it completely covers the brain structures from which it developed. Millions of lines of communication connect one region of grey matter with another and in turn with distant organs such as the lungs, heart and so on. George Gray has said, 'it is the supremely distinctive organ of the human species. What goes on within its network of cells makes the fundamental difference between man and brute. It also distinguishes man from man. It marks the fateful difference between the meek follower and the dynamic leader, between the scholar and the artist, between the genius and the moron – [without] the grey cortex of two billion cells there can be no thought, no sweet sonnets of Shakespeare, no joy and no sorrow.' It is about one tenth of an inch thick and has an area of 400 square inches. This area far exceeds the internal surface area of the containing skull so it is folded and contorted to increase the surface area. It is of great interest to note that the extent of the folding relates to the intelligence of the animal, e.g. in stupid mammals such as the opossum and in birds and reptiles the cortex is smooth and unfolded. Maps of the human brain have been and are being plotted, information coming from the effects of local destruction of brain tissue or by stimulation by electrodes. Some very striking results have been achieved, e.g. in one case stimulation of a certain part of the brain evoked recall of someone speaking or singing with the same speech or melody being repeated exactly each time the particular spot was touched. The sense of recall was not like memory but seemed to be reality, with all the background detail of sight and hearing included, so it looks as if the basis of total recall exists. There are going to be some very exciting developments in 'mind biology' in the future.

Single cells plants and animals

The earliest beings which we would have considered as living probably consisted of a single cell and within it all the living characteristics were concentrated; eating, digesting, reproducing, movement were all carried on within this invisible blob of living matter. Many of these one-celled plants and animals are still with us today – amoeba, for example. These living examples should not however be regarded as the ancestors of more complex forms since they themselves are descendants of the primitive unicellular organisms; but they give us clues as to what their, and our, ancestors looked like.

The separation into 'plant' and 'animal' forms probably took place at a very early period. Some unicellular organisms possessing the green colouring matter, chlorophyll, would be the starting point for the great variety of plants that were to clothe our Earth in the millions of years ahead; others lacking chlorophyll would give rise, in time, to animals ranging from the flea to the elephant.

Reproduction without sex

The method of reproduction of these early unicellular plants and animals was a simple one – they merely split down the middle, the halves separated, and thus we had two where formerly there was but one. Sometimes however the halves did not separate but remained joined together. Similarly at the next division there might again be no separation so that instead of a one-celled organism we now had a four-celled one. As the process continued so we would have multicellular organisms. This obviously had an evolutionary advantage because we find that most living things with very few exceptions – including plants, insects, reptiles, birds, mammals, Man – are all built up of many cells.

With the multicellular organism – whether plant or animal – came what is termed differentiation, i.e. some cells took on certain tasks such as reproduction while others became modified for food digestion and so on. Thus in time a complex living thing evolved with a sharing out of the various tasks within the body.

We do not as yet know what causes certain cells to differentiate – why certain cells for example should become liver cells while others become brain cells and so on. It is one of the major mysteries that is still eluding the biologist though plenty of scientists are involved in its investigation

Sexual reproduction

Sexual reproduction started at an early stage. Instead of reproducing by simply dividing, some plants and animals produced special cells – the sex cells – two of which joined together and the resulting 'zygote', as it is known, then divided again and again to form a new individual. This sexual reproduction was a striking success and it was the pathway that nearly all plant and animals were to follow. Those living things which did not develop it remained low in the evolutionary scale.

Why was it so successful? The answer is that the joining of two cells, each with their own hereditary make-up, led to a completely different individual being produced. At one stroke sexual reproduction increased the variability within a species and natural selection went to work to secure the advancement of the species.

The concept of evolutionary time

One of the concepts that is difficult for us to grasp is the extent of evolutionary time. We talk and write glibly about millions of years but this has no reality in relation to our own life span or even to recorded history. Occasionally a comparison may help. The astronomer, Dr Archie Roy, tells us that if we assume that the universe began on 1 January, the present moment being midnight on 31 December of the same year then (accepting Lemaître's theory) the Earth formed no earlier than November of that year. Life began on our planet about 10 December. About half an hour before midnight on 31 December Man made his appearance. Ten seconds before that last midnight of the year the Sumerian civilization flourished and incidentally the three and a half centuries of telescopic astronomy are crammed into the last third of a second.

Another aid is offered by Dr Stebbins in his book *Process of Organic Evolution.* Suppose, he says, that all of the events of modern history, from the birth of Christ to the present day, the past 2000 years, were compressed into a television show lasting an hour. Suppose then that all of the earlier events in the evolution of life were presented in a series of shows on the same time scale, i.e. each one hour show representing 2000 years. Assuming that these shows were on view 16 hours per day, 7 days per week, then the evolution of Man from Proconsul would require ten months to present and the entire continuous fossil record of living things (600 million years) would require about 60 years. From the beginnings of life the show would take between 300 and 500 years.

Fossils

Most of our evidence concerning the past is derived from the study of fossils. More and more are being discovered and the story is being gradually pieced together. It is a long and painstaking business however because, as we have noted, very few animals and plants are ever fossilized and the vast majority of those that are are locked away for ever in the rocks.

Although many thousands of different kinds of fossils have been discovered they represent only a tiny fraction of all the organisms that have ever lived. Furthermore what we find is a highly biased sample because, in order to be fossilized, the animals or plants have to die under water or on water soaked ground, so that nearly all of the fossils that we find represent organisms that lived in such regions.

By contrast we have practically no fossil record of animals or plants that lived in dry regions or on mountain slopes. This is a great misfortune since it is probable that it was in these very areas that evolutionary change was at its most active, as organisms had been pushed out from their 'safe' areas into more hostile and less highly populated regions where the selection pressures of the environment would be at their greatest. Nevertheless, bearing in mind all these deficiencies and drawbacks, we have a quite remarkable story to tell from the evidence of the rocks.

What is to follow?

In the remainder of this chapter we shall be looking at the evolutionary story of animals without backbones. Starting with very simple little water creatures like hydra we progress through the sea-anemones and jellyfish to the worms and thus to the group that is undoubtedly the most successful of all the non-backbone types. This is the arthropods and it includes crabs, lobsters, spiders, centipedes, millepedes and, most important, the insects which alone comprise about two-thirds of all known kinds of animals. It is not surprising that we dwell on the insects in this chapter. They are the only real competitors of Man and measured by their success they are the 'humans of the invertebrate world'.

Another major group found mainly in the sea but some of which have invaded the land are the molluscs – clams, mussels and snails, squids and the octopus. Both the squid and the octopus have got well-developed brains and eyes but the road towards a backbone and thus towards Man himself lies not via the squids but with some rather insignificant filter feeders such as the sea squirts. Man descended from a sea squirt! The story is an intriguing one.

Sponges

Among the simplest and most primitive of the animals are

the sponges. Fossils have been found in the Lower Cambrian rocks which date back some 600 million years. The simplest of these sponges are vase-shaped, attached at the lower end, with a large central cavity, an opening at the top and numerous small pores in the side walls. There are two layers of cells, the inner lining being made up of cells which have whip-like projections which keep the current of water flowing into the 'vase'. The outer cells form a protective layer.

Although the sponges bear a superficial resemblance to a simple digestive system of a high organism, it is generally agreed that evolutionary-wise they are a dead end. The future of the animal kingdom was not to lie with them. They are one of Nature's less successful experiments although surprisingly, and with unusual compassion, she has allowed them to continue.

Corals, sea-anemones and jellyfish

The road upward was to be via the group comprising corals, the sea-anemones and the jellyfishes. One of the members which is well known to all students of zoology is the little animal, *Hydra*, which is only about half an inch in length when fully extended. It is like a slender vase with eight waving 'arms' at the upper end. The outer cells form a protective layer surrounding a simple gut which is lined with digestive cells. It has also got stinging cells by means of which it is able to paralyse its prey. These cells have coiled up threads which can be shot out rapidly in the presence of food, stinging and injecting a poisonous substance into the body of the animal being attacked. The interest from the evolutionary point of view however lies not with the stinging cells but with the structure of *Hydra*'s body.

Many marine forms are found, some of them formed by budding groups or colonies of *Hydra*-like organisms which may be termed polyps. Some of these polyps have nerve cells scattered through their bodies.

The jellyfish is really just a large polyp turned upside down so that the opening is underneath and the tentacles hang down. Here we find the nervous system beginning to develop. The jellyfish has two interconnecting nerve nets, one controlling swimming movements, the other feeding and other movements. Around the margin of its body it has eight sense organs. Each sense organ has an eyespot sensitive to light, a statocyst which gives information on balance and two sense pits associated with food recognition. Associated with each sense organ there are bunches of nerve cells known as ganglia. Each ganglion has nerve connections with its appropriate sense organ. This system gives the jellyfish the power to react quickly to any situation. Some members of this group live a part of their life as a fixed polyp and part as a free swimming jellyfish. The jellyfish produce sex cells which on fusion give rise to polyps. This 'alternation of generations' occurs in some other animal (and plant) groups. Corals are polyps with a hard coat of chalk for protection.

We have few fossil records of jellyfishes because they are so soft but some impressions have been found in the Pre-Cambrian rocks dating back more than 600 million years. They are obviously among the most primitive of the animals. Many fossil corals have been found in the Ordovician rocks (500 million years ago).

The worms

Whereas the polyps sat and waited for the food to come to them in the water currents their descendants in the evolutionary line went out to forage for it for themselves.

These were the worms and amongst the simplest of these are the flatworms. They are roughly shaped like a spear head and are capable of movement. They have advanced beyond the polyps in that some cells have differentiated to form two nerve cords with a pair of swellings to give a primitive 'brain', primitive eyes and reproductive and excretory systems are also present. We have come rather a long way from the polyp. We have no fossil evidence of intermediate forms but the digestive system is still at the polyp level – the gut may ramify a bit more but still the mouth and the anus are one and the same pore.

Some of the worms such as the marine worm *Nereis* which is about one foot in length are much more highly developed being segmented and having an efficient system of blood vessels, a highly developed nervous system, gills – paddle-like structures for obtaining oxygen – and a fairly complex digestive system, and it has eyes. Many of the intermediate stages between flatworms and *Nereis* can be traced in the worm family.

The arthropods

Moving up the evolutionary scale we come to the crabs and the lobsters, the scorpions and the spiders, the centipedes and the millepedes and the insects all grouped together into the great phylum known as *Arthropoda*. What they all have in common is that they are jointed and Professor Romer (see Further Reading) has said that the best way to get the broad picture of an arthropod is to imagine a progressive worm which has developed a pair of legs on all or nearly all the segments of its body and has then encased himself entirely in armour – armour with well-marked articulations at the joints. They are animals which by mutations have developed skeletons on the outside of their bodies; these skeletons are made of chitin, a horny material similar to our finger nails. The arthropods have a well-developed nervous system with a primitive brain. The blood system is a simple one; the blood after bathing the various organs oozes back into the heart through a series of large blood cavities. The members of this group grow by moulting, the outer skeleton being cast off as the soft body within it expands.

The arthropods have developed from marine worms but they are not of course descended from present day marine worms, which are nearly all highly evolved. The fossil record does not tell us a great deal about links between the two, but we get clues from a little living animal called *Peripatus* which is about two or three inches long and looks like a caterpillar with its short stubby legs. It has features of both the worms and the arthropods. The circulatory system and the head is like that of the arthropods. The legs are not jointed however and it has no hard outside skeleton.

Peripatus however is a land animal and the ancestor that we are looking for must have lived in the sea. Some fossils have been found in Cambrian rocks of animals rather similar to *Peripatus* which were sea dwellers and which thus form a more direct link between the two groups.

The three great groups of *Arthropoda* of the present day are the crustaceans (the shrimps, barnacles, crabs and lobsters which are so abundant in the seas), the arachnids (mites and ticks), and the insects (which dominate the dry land and the lower air).

The first invasion of land – horseshoe crabs and scorpions

At some stage in the Earth's history some of the arthropods moved out of the sea to invade the dry land and the oldest known land forms are the scorpions which make their appearance in Silurian times (440 million years ago). The change from water to land confronts an animal with many problems. One of the most important of these is the danger of drying out. Fortunately from the point of view of our new arrivals on land the hard outside skeleton possessed by the crustaceans in the ocean provided sufficient protection. It also provided body support which would be required when the support of the external water was withdrawn.

Another major problem is breathing – the changeover from extracting oxygen from water to absorbing it from the air. In this however the group that we are considering was well placed. The modern horseshoe crab (which is not a true crab), the only arachnid left in the water, lives wholly in the seas along the Atlantic coast and the Pacific coast of Asia and it breathes by means of gills arranged like the leaves of a book. These gills are situated beneath the abdomen and will still function and take oxygen from air provided they are kept moist. In the scorpions which are land-living, mutations brought the gills from the primitive arachnids into the abdomen where they could be kept moist, connecting to the exterior by narrow openings. The transition from gills to a primitive type of lung was brought about by this rather simple step.

From scorpion-like forms, other arachnids arose – the spiders, the mites and the ticks, but the fossil record is rather thin.

The second invasion of the land – the insects

This however was not the only invasion of the land by arthropods. A second one gave rise to the insects. The insects differ from the arachnids in that they have three pairs of walking legs while arachnids have four, and the body is segmented into head, thorax and abdomen. Almost all forms, except for a few degenerates and a few very primitive types, have wings mounted on the thorax and all of them have 'compound' eyes made of hundreds of little lenses. They are probably derived through marine-living types rather similar to present day centipedes, which live on land, and are segmented and have a pair of little legs on each segment. We have noted that the arachnids got round the difficulty of breathing in air by enclosing their gills within the abdomen thus forming primitive lungs.

In most animals the oxygen extracted in gills or lungs is transported by the blood-stream to the cells in the various parts of the body; these cells in turn give carbon dioxide to the blood which circulates back to the gills or lungs and there gives off the carbon dioxide.

The insects developed another method which is unique to them. Here there is no gill or lung, and there is no dependence on the blood for transport of oxygen and carbon dioxide. Instead little holes or pores are present on the sides of the trunk from which a series of little tubes, termed tracheae, branch out into all parts of the body and carry the air to all the cells of the body.

The importance of the insects

The insects have been very successful in colonizing the land and the lower air. They might indeed be regarded as Man's only real competitors. They compete with him for food and a number have even taken to living off Man. In so doing they may inadvertently inject disease organisms into him which again threaten his very existence.

The number of individual insects alive at any one time is thought to be about 10^{18} (one billion million). Of this vast multitude 99.9% are either innocuous or helpful from the human point of view and a few are indispensable, e.g. bees in pollination. The troublemakers are the other 0.1% – a small percentage which however amounts to about 3000 species. Typhus fever for example, caused by a little organism carried by the louse, has wreaked havoc among armies and civilians alike. Together with plague (the bacteria of which are carried by rat fleas) and dysentery (caused by bacteria transported by the house fly), typhus during the Thirty Years War reduced the population of affected areas from 16 million to 6 million – 10 million killed of which only about 350,000 died in the 'legitimate' slaughter of warfare. Plague alone, or the Black Death as it was called, accounted for more than 25 million deaths in Europe (out of a total population of 100 million) between the years 1345–1351.

Other insects such as mosquitoes, although not actually living in such close personal proximity to Man as the lice and fleas, have nevertheless found him good to eat.

The female mosquito while feeding on Man injects into his blood-stream the parasite of malaria – possibly the greatest killer of all time. It has killed in its time more people than there are people living in the world today i.e. more than 3500 million. About one-third of the human population suffers from malaria.

Among those who have died of the disease in the past have been famous people – Alexander the Great, aged 32, in 323 BC; Oliver Cromwell, aged 59, in 1658; and Lord Byron, aged 36, in 1824. At one time in the Middle Ages the mortality of popes and cardinals was so high that the Papal seat was moved from Rome to Paris. It is said that the lost cities of Ceylon and of Ankar Wat in Cambodia and of the Maya civilization in Central America were deserted owing to malaria. The fall of the Greek and Roman civilizations has been attributed by some historians to malaria. In Rome indeed the population made fever a goddess and sanctuaries were built to her.

Mosquitoes may also inject the virus of yellow fever which left Napoleon with only 3000 men out of the 25,000 he had sent to win a West Indian Empire and which later brought work on the Panama Canal to a virtual standstill.

Insects are also responsible for enormous crop losses both in the field and in store. That classical insect of the Bible, the locust, is still active today, a potential pest over nine million square miles of Southern Europe, Africa and Asia, ranging from Southern Spain and Asia Minor, the whole of northern Africa, through Iran to Bangladesh and India. Within this area more than 300 million people are liable to suffer devastation of their crops.

The locust is a strange insect. For a number of years they may be few in number. Then there are sudden and unpredictable increases in a local population followed by mass irruptions. Swarms may comprise up to 1000 million individuals capable of consuming some 3000 tons of food in a day, destroying everything that is green. A plague may last up to six years and then just as suddenly there is a decline and the few survivors resort to the solitary phase. It is not surprising that Man has invoked the help of insecticides in his attempts to deal with insects.

Flying and non-flying insects

There are about one million species of insects, far more than all other animal and plant species combined, and they occur in a bewildering variety.

Some insects have reached a high degree of organization and have developed great 'skills' as agriculturists. Their crop is a fungus which is a lower form of plant life that lacks the green colouring matter chlorophyll. A number of species of tropical termites make their nests in steeple-like mounds of hardened earth which may be as much as thirty feet in height. The chambers within are ventilated by an elaborate system of vertical conduits extending to the surface of the nest. The nest contains a fat queen and a king enclosed together in a little cell bounded by earth. The smaller workers, soldiers and young termites of various ages and both sexes are also in the nest, at certain seasons with winged male and female future kings and queens. In each nest there may be several fungus gardens. The workers chew and swallow plant debris and pass their faeces on to the garden, as organic manure. On it the fungus grows like a grey sponge. Each fungus garden is enclosed in a close-fitting cavity lined with a mixture of saliva and soil. Only the workers feed on the fungus gardens and pass their resulting faeces back on to the gardens. The king, queen, young termites and soldiers do not eat the fungus but they are fed saliva by the workers. However when winged termites are leaving to found a new nest they take some of the fungus with them to start a new crop. Even more surprising is the fact that during the rainy season the workers may carry some of the fungus out into the open, spread it out and allow it to develop into little toadstools. They then collect spores from these toadstools and carry them back into the nest to start a new garden. It has been suggested that in this way the termites provide for the fertilization of their fungi, as Man does his cereal crops.

Some ants cultivate little animals as the farmer husbands his cows. These have been called pastoral ants. It has been known for more than 200 years that some species of ants visit greenfly (aphids) to take honeydew from them. What is not so well known is that not only do they 'milk' the aphids by stroking them with their antennae but they actually shepherd the aphids to the more tender parts of the plant so that the improved pasture results in a better supply of honeydew from the aphid. If danger threatens the aphids, the ants will pick them up and carry them off to safety. Some even take the aphids underground with them every night and bring them out to graze again in the morning. Others collect aphid eggs in the autumn, take them to their underground nests for overwintering and bring up the newly hatched aphids in the spring.

Communication among certain groups of insects has also been developed to a very high degree. It is now known for example that bees communicate with each other through sound, scent, touch and movement. Much of the information in this field stems from the work of the Austrian zoologist Karl von Frisch who showed that bees visiting a nectar source went back to the hive and by dancing gave very precise instructions to the inmates as to how to get to the nectar, even indicating the direction in relation to the position of the sun and the distance by the number of times that the dance was repeated. Readers wishing to know more about this intriguing story should read 'The Language of the Bees' and 'The Evolution of Bee Language' (see Further Reading).

Typically insects have two pairs of wings mounted on the thorax and they have specially developed muscles to beat them. Some of the larger insects can manage only a few dozen beats per second but the house fly reaches 200 per second and some very small flies may manage as many as 1000! However not all insects fly. Some of the very primitive types, such as the springtails, have got a little forked tail under the abdomen which can be suddenly extended to propel the little insect forward. The silverfish – those lovers of books – also have no wings.

The vast majority of insects have got wings – the dragonflies which may have a wingspan of seven and a half inches (though this is small compared to the two feet wing-span of some of their relatives that thrived 350 million years ago in Carboniferous times), the butterflies with their beautiful and varied colours which are due not to pigments but to the positioning of tiny wax scales on the wings – their beauty indeed will never fade – the bees, the flies, the beetles and many more.

With some insects, such as the grasshoppers, termites and cockroaches, flying is the exception rather than the rule. Others such as the fleas and the lice have lost the power of flight altogether and have reverted to wingless types but they seem nevertheless to be able to compete fairly successfully with any other forms of life.

Life history of insects

In some group of insects the young hatching from the eggs are miniature editions – 'nymphs' – of the adult and they grow to maturity by a series of moults. Others however, such as butterflies, go through an intermediate stage – the larva or caterpillar stage (which is very reminiscent of *Peripatus*) before becoming encased in a hard shell – the pupa – from which the adult insect eventually emerges.

The rocks do not tell us much about the origin of the insects. Many of the ancient insects were large and fossils of them have been found in the coal shales of the Carboniferous period and many insects embedded in amber have been found in the Cretaceous and Tertiary periods (135 million and 70 million years ago respectively). Springtails have been found in Devonian rocks (400 million years ago). The butterflies and moths are not known before the Tertiary period at about the same time as the flowering plants were coming to the fore. Early members of the flies and the mosquitoes are present in the Jurassic period (180 million years) but they became far more numerous in the Tertiary. Bees can be traced back to the Triassic (225 million years) and beetles to the Permian (270 million years). How and when wings arose is a mystery – we do not appear to have found any intermediate forms. Even the origin of the insects themselves is a bit of a mystery. Apparently they did not arise from the crustaceans or the arachnids. Possibly they stemmed from the centipedes but on this the fossil record has little to say.

The molluscs

One major group of animals found mainly in the sea, with a few which ventured on to dry land and even into fresh water, is the *Mollusca.* They too probably evolved from primitive worms but instead of taking the path trodden by the lobsters they settled down to a sedentary existence, evolved a covering called the mantle over their bodies and on top of the mantle a hard chalky shell developed for protection. It has proved to be a very effective protection for the molluscs have survived well. Some of them like the garden slugs and sea-slugs have even dispensed with an external shell though many retain a small internal one.

Among the best known molluscs are the snails which carry their shells around with them. They retreat into these shells only when danger threatens. Gills are found in the marine forms while the land and pond forms have developed sacs that act like lungs. Snail-like fossils go as far back as the Pre-Carboniferous period.

Clams and mussels are also molluscs but in these the shell is hinged so that it can be closed as and when required. They feed by taking water in through one tube and putting it out through another, meanwhile straining it through the gills and retaining any food particles which are transferred to the mouth. We find examples of these animals in Ordovician rocks (500 million years).

The most advanced of the molluscs, such as the squid and the octopus, have well-developed brains which help them in their search for prey. They have given up the sedentary habit. They evolved jet propulsion long before Man thought of it. In the head there is a funnel-shaped aperture which opens into a cavity, which in turn has a narrow tip opening on the body. When movement is needed the whole apparatus is contracted by muscular action, squirting out a stream of water thus propelling the animal with a sharp jerk either backwards or forwards depending upon the direction in which the tip is turned. They also developed tentacles – eight for the octopus, ten for the squid – with adhesive discs which enable them to hold on to their prey. The eyes are also well-developed and in structure they are very similar to human eyes, though it is certain that humans are not descended from the squids. All of this adds up to very efficient, formidable animals.

Most squids are small but the giant sea squid is up to 50 feet long including the tentacles. It is not surprising that they have dispensed with their protective shell. Although the shell in the squid begins to develop in the young it never comes to anything and in the adult it is only a horny plate buried deep in the animal's leathery covering. The squids are elongate in shape whereas the octopuses are rather globular and have no trace of a shell. The relatives of both are found as fossils in the Jurassic period.

Echinoderms

This group includes brittle stars, starfish, sea urchins and sea lilies. One of their distinguishing features is the development of a protective shell which is unlike that of all others we have examined in that it is laid down within the substance of the animal's tissues, thus coming closer to the backboned animals which we will be discussing in the next chapter.

The sea-lilies have each got a stalk with five branched arms leading down to the mouth. The arms have got protective plates on the outside but on the inside of each one there is a groove. Along these grooves on each side there are little soft projections which serve as gills but which also carry food down to the mouth.

The starfish is like a sea lily turned upside down with the stalk removed and it walks on those soft projections which are controlled by a hydraulic pressure system causing them to expand and contract. The starfish cannot feed by filtering and again the tiny feet come to the rescue. It climbs on to a mussel, attaches its tube feet to the shell, pulls until its victim's muscles tire and the shell opens wide enough to allow the starfish to insert its own stomach through the opening and so digest the mussel's flesh.

The sea squirts (Tunicates)

However when we look for the line that was to lead to the backboned animals, and thus to Man, we happen upon little animals consisting of little more than cellulose-containing covering or tunic inside which there is a filtering chamber with a small stomach, intestine and reproductive organs. There are approximately 2000 known species and some of them are attached permanently to the sea-bed. We must note them well; improbable as it may seem they are our ancestors.

Further Reading

J. R. Beerbower *Search for the Past*, New York 1960.
R. Buchsbaum *Animals without Backbones*, London 1948.
H. Esch 'The Evolution of Bee Language', *Scientific American* Offprint No. 1071.
M. F. Glaessner 'Pre-Cambrian Animals', *Scientific American*, March 1961.
A. Krogh 'The Language of the Bees', *Scientific American* Offprint No. 21.
P. A. Moody *Introduction to Evolution.*
A. S. Romer *The Procession of Life*, London 1968.
G. G. Simpson *Life of the Past*, Yale 1953.
James L. B. Smith *The Search Beneath the Sea: The story of the Coelacanth*, 1968.
G. L. Stebbins *Processes of Organic Evolution*, New York 1966.

7. The Procession of Life — Animals with backbones

The vertebrates

In the sea, myriads of fish of all sizes, shapes and colours; on the land, the warm-blooded mammals to be found from the poles to the equator; in the air, the birds; the reptiles that slither through the grass, scurry across the warm stones or wallow in the swamps; and the amphibia that pass part of their lives in the water and part on the land. These are the vertebrates.

What do they have in common? An internal skeleton of bone or cartilage and particularly backbone, the vertebral column, consisting of a joined series of bones running the length of the body, within which is a long hollow nerve—the spinal cord—filled with fluid. Furthermore most have the spinal cord developed to a greater or lesser degree as a brain; but the backbone and the spinal cord are basic to all.

We shall note in this chapter that by assessing the evidence from fossils and from certain living forms, and adding some deductive reasoning, we are able to trace the ancestry of the warm-blooded mammals, including Man, from the sea squirts that we met in the last chapter.

Between these two extremes we shall meet fish that developed lungs and had some of their fins altered to form limbs so that they gave rise to the amphibians. We shall note that some amphibians had their eggs covered with a hard coating, became independent of the water for reproduction, and gave rise to the reptiles; that the scales of some reptiles mutated to feathers and such reptiles becoming warm blooded, gave rise to the birds. Other reptiles, also becoming warm blooded, stayed on land and developed their fertile eggs inside their bodies and gave rise to the mammals.

It seems an improbable story, but it happened.

The tadpole of the sea squirts

For the origin of these backboned animals, fish, frogs, reptiles, birds, mammals (including Man) we have to go back to the little sea squirts which we left in the last chapter firmly attached to the sea bed, filtering their food out of the water. All the backboned animals sprang from these improbable ancestors.

Sitting on the sea floor filtering water is a fairly efficient way of getting food with little expenditure of energy, but when it came to extending its territory (as it would have to do for its numerous offspring) then the fixed sea squirt was at a serious disadvantage. When however some of them by a series of chance mutations developed a motile stage then the difficulty was overcome. This motile stage, or tadpole, was able to swim away, find new territory, settle down and develop into a fixed sea squirt. We know that this happened because it is still happening today. Some sea squirts still produce this motile stage.

Here however is the strange twist. Some of these tadpoles instead of turning into sea squirts continued an independent existence as tadpoles and developed sexual organs prematurely. They were indeed the forerunners of the fish.

The first little primitive 'fish' had come into existence almost as a by-product of the sea squirt! It was as if frog tadpoles, instead of turning into frogs, had developed sexual organs and continued their existence as tadpoles.

A rather inauspicious start for the great vertebrates – a little tadpole with gill slits for breathing in water, a nerve cord running along the back of the body and a notochord, a long tubular structure, the forerunner of the backbone. We note that these notochords can be seen in the embryos of all the higher vertebrates, indicating the ancestry of the spinal column.

Amphioxus – a step towards the fish

There are living today one or two little sea creatures which are not far removed from our ancestral tadpole. One that is well known to most zoology students is Amphioxus – it looks like a tiny translucent minnow but it has no backbone and it has no brain – only a nerve cord running through its back. It also has a well-developed notochord and about 50 pairs of gill slits on its body. Although it can swim, it spends most of its life burrowed tail down in the sand and filters the water through its gills, revealing its filter-feeding ancestry. The group to which Amphioxus belongs is another step towards the true fish.

The lamprey – another step forward

The next step forward has been taken by the lampreys – these are fish which may be two feet in length. Like Amphioxus, they lack jaws and the round mouth with teeth forms an adhesive sucker by means of which the lamprey can attach itself to other fish and prey upon them, but it has developed a three-chambered heart and has a fairly large brain, both characteristics of the fish in general.

The jawless fish – and land plants

In the fossil record we find many of these jawless fish which fed by sucking debris from the bottom of the streams they lived in. They had no fins, swimming by a series of undulating movements. The whole body was encased in an armour of bony plates. These jawless fish were abundant in the water between 500 to 425 million years ago, at a time it may be noted when the first plants were moving from the water on to the land. These early land plants were probably flattened so that they could remain close to any moisture that was present. They were probably not unlike seaweeds which by this time were abundant in the oceans. At first they absorbed water all over their bodies but in time parts of their bodies became specialized for this absorption and for support. We have fossils of some of these land forms going back about 340 million years. We may note that the scorpions were moving on to the land at about the same time as the early plants but it is obvious that there could be no massive invasion of the land by animals before the plants were well established because of the dependence of many of them on plants for food. Even the carnivorous types depended ultimately on plant-eating species.

Fish with jaws

If we now move another 35 million years to the Devonian period we find fish with well-developed jaws and with paired fins. This period is sometimes called the 'Age of Fishes'. The jaws made them much more efficient at catching their prey and the fins gave them much greater control of movement in the water. They were still armour plated.

Cartilaginous and bony fish

However changes were taking place. Mutations that replaced the armour plating by fine scales gave them increased mobility. Fish were becoming dependent on speed rather than armour for survival.

In the early fish the skeleton was made up of cartilage – a soft pliable rather rubbery material and we still have some representatives of such fish living today, e.g. sharks and dogfish. In the majority of fish however the cartilage was replaced by bone. This line gave rise to the wide variety of bony fishes that inhabit the waters of the world today. Water is 800 times denser than air and the oxygen in it is 30 times more dilute than in air, so that fish have to expend considerable energy in order to breath. It is not perhaps surprising then that many fish augment their supply of oxygen by taking gulps of air. The Indian perch has special air chambers which allow it to come out of the water for considerable spells. It comes to the surface to gulp air and will drown if prevented from doing so. The household goldfish will do this too if the aquarium water becomes short of oxygen.

Lung-fish

One group of fish in the face of drought took a path that was to have very far-reaching results. By mutations this group developed a pair of pouched outgrowths from the pharynx which were primitive lungs. They were inflated by air taken in from the mouth so that the fish could survive in dry conditions. Surprisingly the lung-fish line has persisted to the present day in a few localities in Africa, Australia and South America where their lungs still enable them to survive periodic droughts.

One other major development shown by the group is the separation of the heart into two chambers – one to take the oxygenated blood from the lungs and pump it out through the arteries and the other to take in the deoxygenated blood from the veins and pump it to the lungs for replenishing.

Fossil remains tell us that the ancient lung-fishes burrowed in the mud for a period of dormancy during periods of drought, using their fins, some of which were lobed. The fish were almost ready to move on to the land and at this point we tell the rather surprising story of the coelacanth – the 'living fossil'.

The coelacanth

In 1938 off the east coast of South Africa fishermen brought up a strange fish that they had never seen before. Dr Smith of Rhodes University, Grahamston recognized it as a coelacanth – a fish that was thought to have been extinct for 70 million years. It belonged to a group which had flourished during the period when fish were taking their first tentative steps on to the land.

The specimen was however in a rather sad state of decay before Dr Smith got it and despite highly organized searches it was not until 1952 that fishermen brought another living specimen to the surface north of the Mozambique channel. Since then more coelacanths, ranging from 4–5½ feet in length and 70–80lbs in weight, have been recovered and there is little doubt that they are 'missing links', halfway between water and land animals.

They have a double respiratory system, gills and rather primitive lungs. The fins are borne on a scaly stalk protruding from the body and each fin articulates through a single structure just as the limbs of the frog, bird, dog or man hinge on to the humerus of the arm or the femur of the leg. The fin had become a little limb.

The coelacanth confirms what scientists had already deduced from fossils – namely that this group contained some specimens that crept on to the land and became the first amphibians.

The amphibians

The amphibians (of which present day representatives are the frogs) lived part of their lives in the water and part on the land. In particular they had to return to the water to lay their eggs and they could not stand long exposure to dry air.

Apart from the limbs and the lungs which they developed, two more structures grew in importance in these animals. The first of these was the heart which developed three chambers, thus increasing its efficiency as a pumping organ. The atrium divided into two, the right atrium getting blood from the general circulation of the body (venous blood), the left atrium getting blood from the lungs (arterial blood). The ventricle is single and receives its blood from both atria. Secondly, one pair of gills had become modified as openings on each side of the head, covered with a membrane and thus the first ear was in being. This is the forerunner of our eardrum. Hearing was of much greater importance on land than in water.

The general structure of the amphibian brain and nervous system is not unlike that of fish.

Fossil amphibians, found in the Carboniferous period, indicate that some were 8 feet in length, like giant salamanders, and one can picture them floundering around in the swamps, insecure in their tenure of the land. The temperatures of this period were semi-tropical and in the swamps there was a luxuriant growth of plant life. Giant ferns more than 50 feet high and with leaves 6 feet long were everywhere and even taller plants thrust their 'christmas-tree' like branches into the air. It was this luxuriant plant growth that was eventually to be fossilized as our coal measures. Unfortunately the compression was so great that we cannot see fossils in the coal but in the beds of shale (mud turned to stone) above and below the coal measures we find many well-preserved fossils. The living relatives of these plants are the ferns (some of which are still tree-like in New Zealand) and the little 'horsetails' of our ponds and gardens.

The first reptiles

Somewhere during this Carboniferous period some of the amphibians were developing into reptiles which were eventually to displace them and for a period to rule the Earth. To what do we attribute the success of the reptiles?

Sometimes small changes, as for example the stalked fins of the coelacanths, can have great unforeseen consequences. In the case of the reptiles it was the development of a protective coating for their eggs and we have got fossil reptilian eggs from this period. At one stroke this made them practically independent of their watery environment. Eggs could now be laid on dry land without the danger of their drying out. This indeed is the main difference between amphibia and reptiles. By the early Permian period (260 million years ago) the reptiles were abundant, judging by the numbers and variety of fossils found in the rocks and the amphibians, unable to compete, were in steady decline.

Since the reptilian egg was impervious to water, it was impervious to sperm and so fertilization had to be effected by the injection of sperm into the female before the shell formed, unlike the amphibian eggs which were fertilized after being shed from the female. Within the shell the embryo fed on reserve food material – the yolk – and it was surrounded by a membrane – the amnion – which held a small amount of water because the embryo is really an aquatic organism. The situation in fact is very similar to that in mammals including humans although as we shall see, since the mammals developed their young within their bodies they have once more dispensed with the shell.

The end of the Permian period was also the end of the Palaeozoic era and great geological events were taking place. The Earth's crust was being contorted to form great mountain ranges that were subsequently to be eroded down to form hills. On land many of the tree-like ferns of the coal age forests disappeared.

The Age of Reptiles

The Mesozoic era was the extraordinary Age of Reptiles. The reptiles were to populate the earth, the water and the air, and they were to come in all sorts of bizarre shapes and sizes and all because of the egg. At the end of this period of dominance most reptiles were to disappear in a shroud of mystery. The great fern-like trees were on the decline and were being replaced by many of the cone-bearing trees that are common today. The ponds were inhabited by aquatic insects and several kinds of fishes including lung-fishes and lobe-finned fishes which could obtain oxygen from the air. Reptiles were abundant and in particular the group of reptiles known as the dinosaurs.

The dinosaurs

The well named Tyrannosaurus was the largest carnivorous land animal that ever lived. It was 50 feet long and about 20 feet tall as it stood on its hind legs and it weighed about 10 tons. Its forelegs had huge hooked claws and its large, hinged lower jaw must have given it a perpetual gape. Rows of pointed teeth, some of them about a foot long, made this one of the most formidable creatures to have walked the earth.

However not all dinosaurs were carnivorous. Brontosaurus which reached a length of about 80 feet, due to a long neck and a long tail, weighed about 30 tons but lived exclusively on vegetation. Its head was absurdly small for the length of its body and it had a very small brain. This creature and some of its relatives lived in lagoons and swamps since the fossil record shows that its legs would not be able to give full support on dry land.

Stegosaurus, the plated dinosaur, was about 20 feet long, had a double row of projecting plates down the back and spike-like spines on the tail. The skull was very small and its brain was only about the size of a walnut. Other dinosaurs looked like giant armadillos in that they were almost completely armour plated. They were the 'tanks' of this period. Others had huge horns over each eye and a horn on the nose.

For at least 160 million years the dinosaurs were 'lords of creation'. Then 'suddenly' they became extinct. The rocks of 75 million years ago show no fossil dinosaurs and indeed no fossil reptiles other than those of modern types – such as turtles, crocodiles, lizards and snakes. The disappearance of the dinosaurs has been the subject of much speculation among biologists and geologists and a number of reasons have been put forward. Drastic geological changes were taking place. The great mountain chains such as the Rockies, the Alps, the Himalayas and the Andes were being formed. Continents began to take their present shape. Plant life was changing, the climate was becoming colder and the dinosaurs had little ability to control their body temperatures. The few reptiles that live today in cold climates must hibernate and it has been suggested that the dinosaurs, unable to withstand the cold weather and unable to hibernate, died.

Reptiles in the water and in the air

Not all of the reptiles of this golden age were dinosaurs. Some moved into the sea – the plesiosaurs had long necks and bodies 'like a snake stuck through the body of a turtle' and the limbs were like paddles, and the ichthyosaurs were fishlike with no neck and the limbs were like paired fins.

Another group became adapted for flight – pterosaurs or pterodactyls developed membranous wings supported by the fourth finger of the front limb. The first three fingers bore claws by means of which the creatures could cling to the rocks. The bones were hollow and the brain was relatively large for a reptile. The pterosaurs probably used their wings mainly for gliding. Because they were very active it has been deduced that they were possibly warm blooded. Some were no bigger than a sparrow ranging up to types that had a wingspan of 27 feet.

The reptilian bird – Archaeopteryx

The flying reptiles were not however very efficient in their conquest of the air and they were soon to be replaced by a second group of reptiles that took to the air and developed feathers, a modification not seen in the pterosaurs. Fossil feathers have been found from the Jurassic period

(136 million years ago). We have no fossils of links between scales and feathers but there is no doubt that if the fossil birds that have been found had had no feathers they would have been classified as reptiles.

The finding of a 'missing link' was vital and it was found as a fossil in the Jurassic rocks. This find, Archeopteryx as it was subsequently named, was an exciting one in that it showed a close relationship to the reptiles; the rudimentary wings had claws, there were teeth in the mouth and it had a long tail with a row of feathers along each side of the slender chain of vertebrae.

An efficient bird has to be light and powerful. Lightness in birds is achieved by feathers and by hollow bones, and power by the development of large breast muscles attached to an enlarged breast bone.

One of the major developments is the achievement of a fixed body temperature. This permits a high energy output no matter what the external temperatures may be. Reptiles, not having this fixed temperature, become sluggish when the air temperature drops.

The Mesozoic also saw the beginning of the flowering plants that were in time to become the dominant plants on earth. There are about 200,000 species of flowering plants today (the rest of the plant kingdom comprises only about 34,000 species). They have spread to almost every part of the globe and they range in size from a few inches of some herb-like forms to the giant trees. One thing that they all have in common is the flower.

Their rise coincided with the rise of the insects, particularly in the following Cretaceous period which began about 135 million years ago. We have good fossil flowers and seeds from this period. There is no doubt at all but that flowering plants and insects both owe their prominent position to each other because most of the plants are dependent on the insects for fertilization and the insects are dependent on the plants for nectar.

It is of interest to note too that one of the major reasons for the success of this group of plants is that the embryo is retained in the mother plant as a seed until it reaches a fairly advanced stage of development. There is a striking similarity, as we will see, between this system in plants and the mammalian system in animals. The moral would seem to be: for success, protect the embryo.

Although birds were well established by the end of the Mesozoic era it was in the Cainozoic (beginning about 70 million years ago) that they developed into the great variety of species that we know today ranging in size from the humming bird to the ostrich and in colour from the crow to the peacock. Birds, like the mammals, have very high metabolic rates so that they must have efficient circulation. As we noted the venous and arterial blood of the reptile, although at first separate in their atria, are then mixed in the ventricle. This is not good enough for animals requiring high efficiency so that we find that birds and mammals have four chambered hearts. As well as two atria they also, by developing a partition, have two ventricles. Thus the arterial blood is kept from mixing with the venous blood. Bird hearts are larger and stronger than mammalian hearts.

The brain of the typical bird is much larger than that of a reptile of comparable body size. A good brain is required for the complications of flying and the required sectors of the brain seem to be selected for this. In other directions however the bird brain is not particularly well developed – the bird avoids most of its problems by flying away from them! One ability that is highly developed in many of them however is that of migration. One interesting theory about the origin of migration suggests that birds originated when all the continents were together (Pangaea). Birds attempted to reach their homeland to breed each summer as the continents floated apart and this process has continued each year since then as the continents have gradually drifted further and further north.

This Cainozoic period also saw the rise of a group of animals that were to take over all the empty spaces on earth left by the reptiles, and many more besides.

The Age of Mammals

For the origin of this group we have to go back to the beginnings of the Mesozoic period (230 million years ago) where we find fossils of mammal-like reptiles. We classify them as such because of their teeth and bone structures and the positioning of their legs. In this period they were relatively unimportant beings, being overshadowed by the reptiles; but their time was to come.

The characteristic features of mammals are warm bloodedness, hair which serves to conserve the body heat and mammary glands by means of which the young are suckled by the mother. Another striking feature that can be traced through the mammals to Man himself is the massive development of the brain. The first mammals that can be positively identified are from the Jurassic rocks (165 million years ago), and they were very small – about the size of a rat – and they were insect eaters.

Egg-laying mammals

These early forms were probably egg-laying, revealing their reptilian ancestry. Some descendants of this group, the duck-billed platypus and the spiny anteaters, are still to be found in Australia where they escaped the intense competition to be faced by similar egg-laying mammals in the other continents. Probably in very primitive mammals, as in the present day egg-laying mammals, mammary glands had not fully developed. For example, the female duck-billed platypus secretes milk from glands in her belly, which oozes out through slits and the young lick the milk from the fur.

The marsupials

A second group, the marsupials, took a further step forward. The young were retained within the reproductive tract of the mother and secured some nourishment from the yolk sac for a brief period before being born at a very early stage of development. On being born, they made a hazardous journey into a pouch, the marsupium, on the mother's abdomen and attached themselves to the nipple of one of the mammary glands where development was completed.

These early marsupials are represented today by the opossums and kangaroos of Australia. When born, the young of a large kangaroo are less than an inch in length. Those of an opossum are hardly as large as a bee.

For a time the marsupials had some degree of success but, like the egg-laying mammals, they were unable to compete with the more advanced placental mammals. They did flourish for a long period in South America which was cut off from North America. No true placental mammals evolved there as they did in most other parts of the world and in their absence a series of marsupial animals developed paralleling the placental flesh-eaters of other continents with weasel-like, wolf-like and cat-like marsupial forms.

At the end of the Tertiary period connections between North and South America were re-established and these marsupials failed to survive the competition from the placental forms that flooded down from the north. If this land connection had not been re-established South America might have developed a marsupial 'human'!

In Australia the marsupials fared better because no terrestrial placental mammals reached that country until the arrival of Man and placental animals introduced by him. The marsupials developed in great profusion and many of them are still there today – koala bears, even a marsupial mole and the kangaroos which may be looked upon as the equivalent of the placental horses.

The placental mammals

Marsupial and placental mammals differ in one major respect. As we have noted the marsupial young are born when very immature and they have to complete their development in the mother's pouch. The placental mammals, on the other hand, as the name suggests, have developed a special organ, the placenta, through which the young are nourished to a fairly advanced stage whilst still in the body of the mother. This device has been very successful because the placental mammals have spread rapidly over the earth and have replaced the marsupials wherever they have been in competition. Placental mammals of Cretaceous age are known mainly from fossils collected in Mongolia; they were small insect-eaters giving rise in due course to the modern shrews, moles and hedgehogs. Early, relatively unspecialized, insect-eating mammals probably gave rise to all the other orders of placental mammals including Man. The mammals came to dominate the Cainozoic period as the reptiles had dominated the Mesozoic. Few fossils of the early Cainozoic are known but they include many kinds of mammals, some as large as dogs.

Then from 75-60 million years ago fossils became abundant. This was the great era of expansion of the mammals and we find most of the existing orders such as hooved animals and elephant types on the land, whales in the oceans, and bats in the air. By the middle of the Cainozoic period – say 50 million years ago – mammals were thriving everywhere – grazing, swimming, hunting, flying; hooved, clawed animals in profusion. They had expanded to fill the gaps that had been left by the vanished reptiles.

A wonderful picture of this time has been revealed through the finding of a magnificent fossil bed in the Giesel valley of central Germany. This is an evolutionist's paradise, preserved in unbelievable detail, even down to pollen grains from some of the flowers.

Among the trees are fossil redwoods, cypresses, magnolias, rubber trees, palms, figs, cinnamon and mangoes, The animals are well represented too – snails, crayfish, beetles, butterflies, fish, frogs, boa constrictors, lizards, turtles, crocodiles (with their eggs and the embryos still inside them), herons and cranes, an opposum, bats, lemurs, tarsiers, tapirs and ancestors of the horse.

This assemblage of living things is magnificent but odd. What are those tropical trees doing in Germany? Sequoias (now of California and China) growing side by side by rubber trees (now of Brazil). Boa constrictors and crocodiles in Middle Europe? Marsupials which today are found only in Australia and to a lesser extent in South America living in the Giesel valley? What is a tropical forest doing in Germany?

It appears that in the early epochs of the Cainozoic very warm climates were widespread. Mammals were radiating off in all directions and producing innumerable varieties. This was the golden age of the mammal.

In time however great ice-ages were to follow. In the last million years four great ice periods have occurred with ice sheets covering much of the higher latitudes of the earth and glaciers on all the mountains. This, in contrast to the Eocene, was an epoch of widespread extinction, particularly of the mammals. Some families disappeared that had produced species for 40 million years. Horses and camels vanished from the Americas, though they had probably originated there. Elephants once widespread over Europe, Asia, and America died out leaving survivors only in Africa and parts of Asia.

No new major groups emerged during this period but a group of weak, tropical, sparsely haired, clawless animals were swinging in the trees of Africa. They were almost ready to come down from the trees and inherit the Earth. They were the forerunners of Man.

Further Reading

E. H. Colbert 'The Ancestors of Mammals', *Scientific American* Reprint No. 806, March 1949.

E. H. Colbert *The Dinosaur Book*, New York 1951.

E. H. Colbert *Evolution of the Vertebrates*, London 1955.

T. Delevoryas *Plant Diversification*, New York 1966.

J. Millot 'The Coelacanth', *Scientific American* Reprint No. 831, December 1955.
P. A. Moody *Introduction to Evolution*, New York 1964.
A. S. Romer *The Procession of Life*, London 1968.
A. S. Romer *The Vertebrate Story*, Chicago 1945.
H. W. Smith *From Fish to Philosopher*, New York 1961.

8. The Evolution of Man

The distribution of humans

Man has been eminently successful in colonizing the Earth. The present world population is 3500 million people. This number has been reached in a remarkably short time and continues to increase ever faster.

At the dawn of agriculture, about 8000 BC, the world's population has been estimated to have been five million people. By the time of Christ it had increased to 200–300 million due to this development of agriculture and the food that flowed from it. By 1650 the population had increased to 500 million and by the year 1850 it had doubled itself to 1000 million due to the Industrial Revolution and the consequent increase in the standard of living. This second doubling took 200 years compared with the first doubling which took 1500 years. Between 1850 and 1930 the population doubled itself again to 2000 million. By 1975, the population will be 4000 million, and by the end of the century we will be up to 6000 million.

Today human beings are found all over the globe – 225 million in North America, 276 million in South America, 241 million in the USSR, 34 million in Africa, 456 million in Europe, 1990 million in Asia. Where and how did this very, successful mammal, *Homo sapiens* originate?

Man's resemblance to the apes

The human body is constructed on the same general plan as that of other animals and, most notably, the human skeleton is represented by corresponding bones in the skeleton of apes and monkeys – some resemble them so closely in shape and size that it is difficult to distinguish one from the other.

In an attempt to demonstrate the uniqueness of Man, numerous investigations have been carried out to find some parts of the human body that are not present in other primates. All have failed and Dobzhansky, the great American geneticist, has put it well when he says that the differences between Man and apes are quantitative and not qualitative. Not even in the brain could any unique areas be found.

The human brain is about twelve times as large and much more complicated than the monkey brain but most of this increase in size is concerned with memory, association and speech. So far as his method of seeing the world around him is concerned, Man is still more or less at the stage reached by monkeys 25 million years ago. He smells, tastes, hears, touches and sees as monkeys do.

In addition Man shares with the apes a great many very specialized features; they too have the power of seeing things stereoscopically or in '3D'. This is useful to primates in the trees and it was later to allow Man to manipulate fine tools. Man and apes both have colour vision; the females have a menstrual cycle, an absence of breeding seasons, birth usually of only one offspring at a time which is greatly cared for by the mother, and the male usually has a dominant position in the family or the group.

All the embryos of the backboned animals – lizards, fish, birds, monkeys and Man resemble each other at an early stage of development. The embryos of Man and apes are almost indistinguishable up to a fairly late stage in development – the human embryo even having a tail before it is absorbed. Man's ancestry goes back through the thousands of millions of years to the very beginnings of life itself. All animals and all plants start from the one common origin. The higher we go up the evolutionary tree the more closely are the animals related to us.

The primates

Mammals came into being some 60-70 million years ago. The order to which Man belongs is called the primates which has about 200 living species, including lemurs, monkeys and apes, all with a large braincase, standing erect, and usually having a thumb and sometimes a big toe which is positioned for gripping. They usually have nails instead of the claws of so many other mammals and they have jaws positioned under the braincase. All of them share the ability to climb by grasping and this ability lies at the root of the whole primate order and makes possible the evolution of Man. They came into existence in the trees – only Man and the baboons have deserted this way of life – and by wrapping their fingers and toes around branches instead of driving their claws into them, as other animals did, they gained mastery of the trees. In their physiology and biochemistry too the primates have a great many features in common.

All of the living primates other than Man are tropical animals and they include (1) the great apes such as the chimpanzees and the gorillas of Africa, the orang-utans and the gibbons of South East Asia, which are all specialized for life in the tropical forests with long arms well adapted for swinging from the branches of the trees (2) the 130 or so species of monkeys of the Old World and the New World, which are smaller than apes or Man, have long tails and mostly live in the trees, though the baboons live in treeless country; they walk on all four legs; they have large and movable ears and their eyes are directed sideways; they do not have colour vision.

It should be clearly noted that Man is not descended from any of these modern day apes and monkeys any more than the reader is descended from a distant cousin. Just as you and your distant relatives have ancestors in common so monkeys, apes and Man have common ancestors not too far back in the evolutionary line.

The question is, how far back and to which of the monkeys or apes is Man most closely related. The question is a fairly open one and the details are constantly changing as more and more discoveries are made. Present evidence suggests he is more closely related to the apes.

Proconsul

The common ancestor of both apes and Man is thought by many to be Proconsul, an animal represented by the remains of some 450 individuals dating back some 25 million years, to a time when monkeys, apes and Man had yet to appear. Not surprisingly Proconsul showed characteristics of all three. He was probably the forerunner of monkeys and apes. The question of his relationship to Man is still open.

The first remains were found by Dr Leakey on a small island in Lake Victoria, Kenya, in the 1930s. They were sent to England where their resemblance to chimpanzees was recognized and indeed the name 'Proconsul' was a sort of private joke – 'pro' meaning before and 'consul' was the name of a well-known chimpanzee in the London Zoo. Proconsul has a skull of both apelike and Manlike characteristics. It is thought that some of his descendants became completely tree-living from which the apes descended, while others may have developed a more erect posture, inhabited the ground and eventually through a long series of mutations gave rise to Man.

Ramapithecus

Another possible ancestor of Man is called *Ramapithecus*. His fossils were found in India in 1934 and similar specimens have subsequently been found in Africa and Europe. The jaws and teeth were very close to those of Man and he lived about fourteen million years ago.

The Abominable Coal Man

Another early primate which must be considered among

the candidates for early Man, rather irreverently known as the Abominable Coal Man, was first discovered about 100 years ago and a nearly complete skeleton was found in a coal-mine in Italy in 1958. This pre-Man was about four feet tall and probably weighed about 80lbs. Studies of its limb bones suggest that it had an upright posture.

Australopithecus

In 1924 a small fossil skull was found in South Africa and it was named *Australopithecus africanus*. Since then many additional fossil remains have been found including ribs, vertebrae, limb bones, skulls and pelvis.

A very complete picture of this pro-Man has been pieced together. He was probably smaller than most humans and weighed about 90lbs. Undoubtedly he walked erect. His kind of upright walking was not however the same as modern man's heel and toe striding gait. It was rather a jog-trot, covering the ground with quick rather short steps with the knees and hips slightly bent. Dr John Napier, an expert in this field, has said that it called for a high output of energy and long distance travel was probably impossible. Many believe that he is the earliest member of Man's family yet discovered. His age is not yet certain but possibly he thrived about two and a half million years ago. This means that there is a gap of more than ten million years between *Ramapithecus* and *Australopithecus* during which the rocks tell us absolutely nothing about Man's ancestry and this gap has been a constant source of irritation to anthropologists.

In many ways *Australopithecus* resembled Man, yet in others he resembled the ape. The shape of the pelvis indicates that these creatures walked erect and this is further supported by the shape of some of the bones in the feet. The brain size however was small and this led some people to classify them as apes. Recent findings however indicate that these creatures were not only tool-users but also toolmakers suggested to many that they were close to 'crossing the Rubicon' between animality and humanity and indicates, as Darwin thought, that Africa was the cradle of humanity.

A new skull is found

The finding of a different skull – as yet it has no name – in November 1972 by Dr Richard Leakey throws doubt however upon the possibility of modern Man, *Homo sapiens*, being descended from *Australopithecus*. This newly discovered skull found at Lake Rudolph, Kenya, is more than two and a half million years old and it does not fit any of the already established categories. The significance of the find is that it is at least a million years older than any previous human skull. This means that it must have co-existed with *Australopithecus*. The great similarity of the new skull to the skull of modern Man (though having only about half the capacity) makes it unlikely that the apelike *Australopithecus* gave rise to Man. Leakey believes that it is much more likely that the new skull is the real ancestor of man and that *Australopithecus* was an evolutionary dead-end which died out about 700,000 years ago.

Homo erectus

There is no doubt however that when we come to some of the finds that have been made in eastern Asia, Java and northern China that we are dealing with Man. Peking Man and Java Man have been grouped into a single species, *Homo erectus*, who lived about 500,000 years ago. Fossils have been found in widely separated parts of the world and there is quite an abundance of material.

Homo erectus walked erect and indeed the femur of his leg cannot be distinguished from the femur of modern Man and he had developed the modern heel and toe gait and probably stood about five feet high. The forehead was low, sloping back from a heavy ridge above the eyes, and the jaws heavy and protruding but the lower jaw had no chin.

His brain capacity was larger than that of *Australopithecus* but much smaller than that of modern Man. He used tools but, most exciting of all, he had discovered the use of fire! Remains of hearths are abundant in the caves in which *Homo erectus* lived.

We now come to rather a thin period in our knowledge of the development of Man. The next 200-300,000 years are represented by only two broken skulls. Both of these are from Europe and they show an increase in brain size together with a higher forehead.

Neanderthal Man

Neanderthal Man appeared in Europe, Asia and Africa about 100,000 years ago, just before the last glacial period. He lived in a harsh environment and thus he took to caves for protection. He appears to have been a strong, stocky build and many of his remains have been found throughout the caves of Europe. He was a hunter and was well acquainted with the use of fire. He dressed animal skins and probably used them for clothing to protect him against the cold and wet. He had a large head and had as large a brain as modern Man. Indeed it has been said that if we put him in a suit of clothes and sent him down to the supermarket for groceries he might pass completely unnoticed. There is evidence also that he buried his dead and this has been interpreted as indicating the first stirrings of a social and religious sense. With the dead he often buried tools.

Homo sapiens

Then some 30–40,000 years ago Neanderthal Man vanished from the scene, perhaps because he was unable to withstand the fierce cold which prevailed in Europe at that time; yet he also vanished from areas not subject to the Ice Age. Some anthropologists consider that he evolved into modern Man – *Homo sapiens*; others that he was exterminated by *Homo sapiens* who appeared on the scene then.

An early form of *Homo sapiens*, appearing some 35–40,000 years ago, was Cro-Magnon Man who, judging by his culture which was quite different from that of Neanderthal Man, was not a native of Europe. His remains indicate that he was very variable in height but some were up to about six feet. He was a great wanderer and in a remarkably short space of time, if we are to judge by his fossil remains, he appeared in Asia, Australia, Africa and Europe. He was a great hunter and he made well-fashioned tools of stone both for killing and for preparing his meats and skins. Some were also artists, indicating a level of culture high enough to draw the beautiful animals on the cave walls of France and Spain.

About 10,000 years ago, still during the time of Cro-Magnon Man, many large mammals became extinct all over the world. The reasons for this are not known but it has been suggested that this disappearance led to Cro-Magnon Man giving up his cave existence and his dependence upon hunting to develop a more village type of existence where he cultivated crops and animals, and was in fact a farmer. He was virtually indistinguishable from people of the present day.

One of the problems that still remains wide open is whether all the present day races of mankind developed from a single stock or whether they are the result of derivation from different forms.

Modern theory on the evolution of Man

Assessment of the course of Man's evolution has undergone a radical revision in the past 30 years or so. Up until the 1940s it was assumed by anthropologists that the spectacular development of the brain came first and led the pre-Man to move down from the trees, to develop an upright gait, to make him a tool-making and tool-using animal. This is not however borne out by the evidence of the fossils. In 1948 Professor Edinger of Harvard University showed that the growth of the brain *followed* the physical development and change; that the development of the body came first and that the development of the brain came later; that we had human bodies before we had human intelligence. Anthropologists elsewhere, and to their surprise, found human bones, in particular human pelvic bones, alongside apelike skulls. There was only one conclusion. 'Man' was walking while his brain was still that of an ape.

About a million years ago, the theory now runs, a change took place that was of fundamental importance. Some of the ground-living apes who had come down from the trees to have access to a wider range of foodstuffs were born, due to mutations, with a different kind of pelvis. No doubt this type of mutation occurred among the tree-living apes but in the trees it had no evolutionary advantage and therefore no matter how often it appeared it would not survive. On the ground however it was a very different matter because this change in the pelvis meant that not only could the ape walk upright but it *had* to walk upright, on two legs. For the first time the upper limbs were free and thus tools could be carried in the hands.

With the aid of tools these upright apes gained an advantage over their relatives on all fours and thus they survived. Not only that but natural selection was off on a new tack. A premium was now placed on brain rather than muscle and so those that by chance mutation developed better brains and were therefore better tool-users were selected again and again. It thus appeared to Professor Washburn of Chicago, who put forward the theory that the whole vital evolutionary cycle began when some of the big ground-dwelling apes were born with a shorter pelvis. It was only after the trunk and the pelvis and the legs had developed much as they are today that the brain then developed until it reached its present stage.

As the new species emerged so he began to move over the various territories of the Earth. His upright stance and his heel-to-toe walk widened his range of activities. At first a vegetarian he learned as he moved out of his tropical forest to eat meat and the tools and weapons that he developed as a result of having his upper limbs free enabled him to ensure an adequate supply of food.

One result of his changing to meat-eater meant that he could not only extend his territorial range but he also had more time for other activities. One good feed of meat would last him for a whole day. As he learned to cook and dry his meat so he found that he could carry it with him and thus survive over fairly long distances. As he reached the colder climes he used skins and furs for protection.

By natural selection he now had a large brain and he was able to invent more complex tools and devices – spears for throwing, bows and arrows, bridges and boats for crossing rivers. As the population increased so the more adventuresome had to move out into unexplored territories. Many would perish but natural selection favoured those with the physical and mental capacity to survive.

So these hunters moved throughout the Old World, out into Australia, up into the Far North across the Bering Straits into Canada and down the length of the Americas to the southernmost tip of South America, leaving behind them as they went those who were content with a more settled existence. So in time the Earth came to be populated by human beings.

Then some 10,000 years ago man ceased to be a hunter. He settled down to cultivate crops and animals. It was the dawn of agriculture and groups became less nomadic and more separated and by new mutations and continuing natural selection, they developed their own characteristics.

The races of Man

Thus Man emerged as a species and began to separate into races at one and the same time but this process was probably accentuated as he became less nomadic.

The following races based on such features as hair

texture, width of nose, and colour of skin, have been recognized by anthropologists. Caucasoid – wavy hair, narrow nose, 'white' skin; Negroid – woolly hair, broad nose, 'black' skin; Mongoloid – straight hair, moderately broad nose, yellow-brown skin; Australoid – curly hair, moderately broad nose, 'brown' skin.

Until recent years some anthropologists have tried to insist that races are separate species, or even separate genera, with Caucasians descended from chimpanzees, Negroids from gorillas, and Mongoloids from orang-utans. Franz Weidenreich during the 1940s claimed that races are descended separately not from such divergent parents as the several great apes but from the less separated lines of fossil-men, for example Peking Man led to the Mongoloids, Rhodesian man to the Africans, and so on.

Such theories are no longer acceptable in the light of modern findings and it is generally agreed that the races developed from a stock showing all the features of *Homo sapiens*.

How do we explain the different facial and other characteristics that are apparent in the different races? If we accept Darwinism as applying to Man then these features must have or have had survival value since they were chosen, from random mutations, by natural selection.

In his fascinating article on 'The Distribution of Man' William Howells (see Further Reading) attempts to answer some of these questions. There is a tendency among warm-blooded animals of the same species to be larger in colder parts of the territory. As an animal gets larger its inner bulk increases faster than its outer surface so that more heat is retained. Secondly, the size of extremities such as limbs, ears, noses, is smaller in colder climes and larger in warmer for the same reason. Man, in common with all other animals, obeys this general rule, there being lanky long-limbed people in hot deserts and dumpy short-limbed people in the Arctic.

The dark skin of Negroes is due to a pigment called melanin which all humans possess to a greater or lesser degree. It screens the skin from the ultra-violet portion of sunlight. The Negro skin has probably been selected in response to the strong sunlight of the equatorial regions, although other suggestions, such as giving protective colouration in the forest, have also been put forward. The woolly hair of the Negro provides a very effective protection against the heat from the sun.

The Caucasian race extends from cloudy northern Europe, where the ultra-violet from the sun is not only acceptable but desirable, down to the intensely hot Sahara. It is therefore of great interest to note that as one progresses southwards and there is therefore a greater need for protection from ultra-violet then the pigment deepens. The Scandinavians are very light skinned with fair hair, the Mediterranean people are swarthy, brownish in Arabia, and dark brown in India. It is tempting to assume that here again we have natural selection at work.

The slant eyes of the Mongoloid are deeply set in protective fat-lined lids, the nose and forehead is flattened and the cheeks are broad and padded with fat. This, says Dr Howells, is an ideal mask to protect eyes, nose and sinuses against bitterly cold weather. This face evolved under intense natural selection some time during the last glacial advance among peoples trapped north of a ring of mountain glaciers and subjected to fierce cold which would have weeded out the less adapted, in Darwinian fashion. This must have happened not so long ago because the population must have been well advanced in hunting skills to survive in such intense cold and, furthermore, the change must have occurred *after* the American Indians, who are Mongoloid but have not got this transformed face, migrated across the Bering Strait. Only the Eskimos in America, representing a later 'invasion', have got the typical eyelids, foreheads and cheeks. Dr Howells therefore considers that the selection took place between 25,000 and 10,000 BC.

The term 'race' is now becoming difficult to define. Its boundaries are becoming blurred. The members of the various human populations are no longer isolated behind mountains or by stretches of sea. The aeroplane and indeed all forms of travel are breaking down the barriers and marriages between persons of different populations are taking place. The United States has been described as the great 'melting pot' where peoples of all races are intermingling. The same is also true of many of its near neighbours in South America. In time we may see *Homo sapiens* returning to a homogeneous condition though not the homogeneous condition in which he set out so long ago to roam the plains of the world.

Man as a social animal

We do not know when Man became a social animal. Possibly even before he became Man he hunted and lived in groups but the specialization within these groups would be small. The males would be hunters and the females would rear the young.

With the coming of agriculture however some 10,000 years ago, leading to a greater availability of food and a more stable existence, specialization would rapidly follow. Thus some members of the tribe, being particularly good at growing plants, would become farmers; others gaining a knowledge of poisonous and drug plants would become witch doctors and medicine men. Some, because of greater muscular or brain power, would become the leaders of the tribe; they would employ servants to do their manual tasks for them and thus a hierarchial system would develop. Some of the servants would be employed as soldiers.

With the coming of the industrial revolution to the western world in the eighteenth century the pace of specialization accelerated as more men were required to guide the activities of the various machines. Some people acquired great wealth and therefore power. They employed managers to oversee the running of their great factories. The manager in turn employed specialists in all the various branches of an industrial concern. Thus our modern society came into being. Man had come a long way from Proconsul.

Further Reading

Theodosius Dobzhansky *Mankind Evolving*, Yale 1962.
S. Eimerl and I. De Vore *The Primates*, Life Nature Library, 1966.
G. W. Gray 'The Great Ravelled Knot', *Scientific American*, October 1948.
F. C. Howell *Early Man*, Life Nature Library, 1966.
W. W. Howells 'The Distribution of Man', *Scientific American*, September 1960.
Ruth Moore 'Evolution revised', *New Treasury of Science*, London 1960.
Desmond Morris *The Naked Ape*, London 1967.
'Human Variations and Origins', Readings from *Scientific American*, London 1967.

9. The Way Ahead

Predictions

One of the most difficult exercises in which one can dabble is to attempt to predict the future. In 1937 a group of top American scientists sat down together and recorded their predictions as to what might be the main technological developments by 1967. Among other things they completely missed computers, radar, most contemporary electronic devices, antibiotics and the jet engine.

Professor John Lenihan, a Glasgow scientist, tells us that Lord Kelvin wrote in 1896 to Colonel Baden-Powell, 'I have not the smallest molecule of faith in aerial navigation other than by balloon rig,' and Newcomb the American astronomer observed a few years later, 'No possible combination of known substances, known forms of machinery and known forms of force can be united in a practical machinery which man shall fly long distances.' In 1956 the newly appointed Astronomer Royal described space travel as 'bilge, utter bilge'.

Prediction seems to be a very chancy business and yet lots of people indulge in it.

Already there is in being a Commission on the Year 2000, an Institute for the Future and a World Future Society. Many people seem to be willing to take a chance. Why shouldn't we?

The population crisis

One of the major problems that is facing Man is the exploding birth rate all over the world. In many of the underdeveloped countries mortality among babies and indeed among the population generally was high because of the twin evils of disease and starvation. Now however with the development of modern medicines such as penicillin, streptomycin and the anti-malarial drugs, with the control of disease-carrying insects due to the use of DDT and other insecticides, with increased food production due to modern agricultural methods, the rate of mortality has fallen sharply. People are living longer and more babies are surviving into adulthood. In India alone DDT helped to reduce the incidence of malaria from 75 million to 7 million in a decade. In Ceylon, following a massive DDT campaign, the number of malaria cases dropped from 2.8 million in 1946 to 110 in 1961 and the number of deaths dropped from 12,587 to zero over the same period. In Venezuela the number of malarial patients fell from 817,115 in 1943 to 800 in 1958. All over the world a similar story is told not only for malaria but also for many other insect-borne diseases such as typhus fever and dysentery.

Although much has been written about the underdeveloped countries the position in many of the advanced technological countries is striking too. In Britain, for example, the population is rising at the rate of 0.6% per year. This is one of the lowest rates of increase in the world as can be seen from the following table (which is devised from Ehrlich and Ehrlich, *Population, Resources, Environment*).

POPULATION GROWTH

	Present Popln. (Millions) (mid 1969)	Present %age increase	Number of yrs to double (projected)
WORLD	3551	1.9%	37
AFRICA	344	2.4%	28
ASIA	1990	2.0%	35
N. AMERICA	225	1.1%	63
LATIN AMERICA	276	2.9%	24
EUROPE	456	0.8%	88
USSR	241	1.0%	70
OCEANIA	19	1.9%	41

but even this low 0.6% increase means a doubling of the population in 100 years. If this rate continued for the next 2000 years it would multiply the population a million times. It would mean that there would be 150 people per square yard. Packed into their multi-storied tenements like sardines they would be emitting so much heat that it could not be radiated into space and they would frizzle to death.

The rate of increase in Asia is remarkable. With a present population of nearly 2000 million and a 2% increase there is going to be a population of 4000 million (more than the present *world* population) by the year 2000.

Already two-thirds of the world's population is undernourished. What is going to happen when the population explosion really gets under way? Famine, disease, war or all three? The Ehrlichs tell us that time is running out on us. If the net reproduction rate in the United States had been reduced to unity in 1965 the population would continue to grow for another seventy years and reach a figure 38% above that of 1965 before stabilizing. The urgency of the problem of achieving a reduction in growth can therefore be appreciated.

On a global scale the present rates of population growth imply a doubling of human numbers in less than 50 years, a quadrupling in less than 100, an eight-fold increase in less than 150 years.

The food crisis

We should not assume that the 'blame' lies solely with the developing nations. Population growth in the better off countries over the next 30 years is projected at 500 million. By present standards each of these people will consume six times as much food as that eaten by individuals in the poorer countries so that the actual effect of 500 million new 'affluents' on Earth is approximately equal to the impact of nearly 3000 million additional human beings being expected in the developing countries by the end of this century. By the year 2000 food supplies will have to be expanded by 160% for Africa, 240% for Latin America, and 200% for East Asia merely to provide a minimum adequate diet for their peoples.

Very significant increases in grain yield are possible. Due to the introduction of high yielding varieties and an increased use of fertilizers leading to the 'Green Revolution' the Indian wheat harvest in 1968 was 35% above the previous record and the Pakistani crop was 37% higher than any previous year. Due to an early maturing of these new varieties and their independence of day length two or three crops may be taken in a year. In the Mysore State of India farmers are growing three corn crops every fourteen months. When adequate water supplies are available Indian, Indonesian and Philippine farmers can grow two or three crops of rice in a year.

It is a great pity that such massive increases are being used up by an ever increasing population rather than alleviating the hunger of a stabilized population. No doubt many new food production techniques will be developed in the years ahead for already the foundations are laid – extraction of protein directly from plants of all kinds including grass; the growing of algae in tanks to give massive yields; the growing of yeast and other micro-organisms as food, on various waste by-products of e.g. the oil industry; and the vast resources of the oceans have as yet been almost untouched. The fisherman is still at the stage that Neolithic Man was at 10,000 years ago when he went out searching for his food. Marine culture must soon follow in the path of agriculture.

Despite all of the new developments and techniques however world population will continue to keep ahead of food production and as the years go on the gap will not narrow; it will widen.

The resources crisis

With increasing numbers demanding (rightly) higher standards of living so our natural resources are tending to run out. It has been estimated that there will be little crude oil and natural gas by the end of the century. Manganese, molybdenum, cobalt and aluminium will have disappeared by the year 2200; other raw materials are dwindling.

Population control

What can be done about these crises? To most thinking people it seems clear that there is only one answer – limit the population by birth control. Man has temporarily removed most of the natural selection processes – disease, famine and so on that in the past limited his numbers and still today limit the growth of animal populations, but he has not substituted anything in their place. He is like a small boy who, limited in the amount of sweets that he can eat by the amount of money that he has available, suddenly finds that all the brakes are off – the sweet shop doors are open and he can eat as many sweets as he likes. Like the small boy who can either eat until he is ill or become self-disciplined and enjoy sweets within reason and remain healthy, so Man has a choice too. He can go on reproducing until the race becomes mortally ill due to overcrowding, lack of food, disease or he can develop a self-discipline, decide to limit his numbers and have a good life for all.

Under such circumstances it is not surprising that in many scientific areas there is massive research going on for birth control measures. The 'pill' is as yet in its infancy. Soon we may have male pills which will prevent sperm production or render such sperm innocuous. Another is the 'after' pill which will destroy any fertilized ova after intercourse. The pros and cons of abortion have also been widely debated. It would seem however that with the eventual development of birth control measures then abortion will in the main become a non-issue.

The pollution crisis

Another of the major hazards that Mankind is facing and

will face increasingly in the future is pollution. This is not a new phenomenon. When Dr Johnson visited Edinburgh he remarked, 'There is a strange effluvium in this city', and Boswell explained, 'Sir, it is from the open sewers.' When the housewives of our cities in the Middle Ages threw up their windows and yelled 'Gardez-loo' before tipping the human excreta to the streets below and, incidentally, on to the heads of passers-by who happened to be hard of hearing or not nimble of foot, they were polluting the environment more surely and directly than their present day successors do around the beaches of our country. Our 'Gardez-loo' cities were not exactly conservationists' dreams but note this: they were accepted as normal until the sewage technologists stepped in to solve the problem.

Increasingly the people of this world of ours are demanding more of the products of our technology. In the United States the amount of goods and services has grown as much since 1950 as it grew in the entire period from the landing of the Pilgrim Fathers in 1620 up to 1950. Technological development has grown as much in the past 20 years as it grew in the previous 320. The United States is *increasing* its annual output of tin cans, plastic containers, electric power, cars, chemicals and other manufactured goods by 300,000 million dollars per year – not only is there a population explosion; there is a corresponding technological explosion too.

As we reap the benefits of technology, and there are many, so we have to cope with its by-products. Industrial waste, oil spills, motor-car exhausts, smoke, plastic containers, noise and our own sewage. There are some scientists who predict that Man will disappear in a sea of pollution – that he will not long have to worry about the population explosion – the pollution explosion will do the job for him if we do not halt technology. I do not believe that halting technology is the answer. This is the best environment in every way, social, cultural, aesthetic and medical that Man has ever lived in in all the years of his existence.

As has been said by Alexander Szalai in the journal *Ceres*, 'Let us not forget that Man in his "natural state" had an average life expectancy of less than 40 years, was mercilessly exposed to hunger, cold and all sorts of "natural" illnesses and ravages; life was poor, nasty, brutish and short. Before the beginning of the seventeenth century twelve famines swept through England every 100 years. It is only in the highest industrialized societies and only there that man can now expect on the average to live 70 years or more and that millions – though as yet by far not enough millions – can enjoy welfare and well-being unattainable on any lower level of economic and technological development.

There is a threat of pollution and in parts of some countries such as the United States of America and Japan it is more than a threat but the technology that creates so many problems will also, I believe, be instrumental in solving the problems. The answer is certainly not to halt technology for technology gives Man many of the good things of life. The technologist however must learn to look at the whole process, not just part of it – the supply of raw materials, the finished product, the waste products of the process and perhaps above all the recycling of materials so that by-products (pollutants) are cut to a minimum and the strain on natural resources is lessened.

If Man survives the population explosion, the resources crisis and the pollution threat as I believe by means of his common sense he will, then what lies ahead?

The control of ageing

For centuries men of power and influence feeling their abilities slipping from them as they aged have attempted to arrest the ageing process. They have searched for the elixir of youth but monuments of stone testify to the futility of their quest. They were doomed to failure because their researches were compounded of folklore and witch doctor mumbo-jumbo from rabbits paws to monkey extracts.

All living creatures seem to have their allotted span –

150 years for a tortoise, a day for a may fly, three score years and ten for Man. Is their life span immutable? Is it predetermined? Must we accept it? Or is it a wearing out process that can be arrested? Much research is being devoted to the subject.

It seems clear that the ageing process is linked closely with the death of individual cells in the body that fail to be replaced – nerve cells, muscle cells, kidney cells. Why should the cells die? Is it as has been suggested by some biologists that the cells produce substances which in time 'clog up' the mechanism of the cell? If this is so then it should be possible to devise chemicals which bring about declogging. If this is so then some believe that this elixir would not only delay ageing but might even stop death and humans would achieve immortality.

Other scientists believe that cells are 'programmed' just as a computer is programmed, to function for a certain length of time and then the programme runs out. If this is so then it is reasonable to think that some day scientists may be able to tinker with the programme and perhaps put it on an endless belt system. The programme, if programme there be, is probably encoded in the DNA of the nucleus.

Immortality may not be a wild pipe dream. Some day scientists will decode the process of ageing just as they have solved and are in process of solving other biological 'mysteries' such as how hereditary characteristics are passed on from one generation to another through an ovum and a sperm which are too small to be seen with the naked eye.

The social consequences of the elucidating of the ageing process would, of course, be enormous. There would be no death, other than that caused by accident or illness and there would be therefore little renewal of life. Would there be a pact by means of which people would agree that they wouldn't have children or would Governments legislate that they mustn't? Would the ultimate step of sterilization be taken?

Even in the case of accidents or illness in the future it may be possible to replace the injured or diseased organ whether it be lung, heart, kidney, liver or limb with a brand new one. At present some such replacements can be effected with varying measures of success but all require a donor. It may in the future be possible to grow new organs from a few cells taken from the original organ.

Plant, animal and human cells from various tissues can at present be kept alive and growing for an indefinite period. In the case of plants these cells by special hormone treatment can be induced to form roots and shoots. At present we cannot make animal or human tissues form new organs but it could be a matter of time and technique. Every cell carries the potential information to enable it to form an organ.

We must not underestimate the difficulties that are involved but if organ-growing became a reality, as I believe it will, then many of the problems of transplantation of organs would be solved. There would certainly be no rejection problems since the new organ would have the same genetic make-up as the organ it was replacing. New lamps for old indeed!

It would appear however that there would be almost insuperable problems in brain regeneration or brain transplantation and some among those who believe that immortality is at least a possibility have expressed the fear that although the body may be kept physically young the brain and the mind may degenerate so that it would be a senile mind in a young body. But need this be so?

Some of the most exciting scientific work of the future may be in the area of mind-biology and it is certainly true that it is becoming increasingly possible, by means of drugs, to affect and alter moods and emotions. Also research into the nature of memory is throwing up some very interesting results suggesting that it may have a chemical basis. If this is so then rejuvenation of memory could be a very real prospect.

There is certainly much work being done on the problem

of ageing. In the United States there are more than 1000 teams working on it and it would be a fair guess that there is at least a comparable number of teams working in the USSR.

Cloning

The normal method of reproduction of living things is that a male seed or sperm fertilizes a female egg or ovum and from this fertilized ovum a new individual develops by cell divisions.

There are other ways, particularly in the plant kingdom. We can take cuttings from say a carnation and produce lots of new carnation plants which are identical to the original. A slightly more refined method of propagating in this way is to separate out a number of body cells (note *not* reproductive cells) and by developing these cells under the correct conditions to grow from each one a fully mature plant.

Such a group of plants all genetically identical is known as a 'clone' and the question arises – will it in the future be possible to clone people?

We will certainly not be able to take cuttings. Arms, or legs, or hands, or fingers just don't grow into human beings. There are however many biologists who believe that it might be possible to take a number of, say, skin cells and grow each cell, by division, into a human being.

At present it is possible to grow such skin cells in test tubes but they just produce more cells. The idea proposed is that by giving them the correct nutrients, vitamins and hormones such cells might be induced to divide and differentiate to form the multifarious tissues and organs that go to make up a human being. A pipe dream? Perhaps, but many biologists regard it as a real possibility; but why do it?

H. J. Muller the famous geneticist believes that the human race is deteriorating genetically and is therefore doomed unless something is done to check the process. Cloning would be one way of stabilizing the human race if we chose our stock wisely. We could clone, for example, from people with high intellects (Einstein would be an obvious choice, if he weren't dead). By means of a cloning process we could have dozens of Einsteins and we could have Einsteins of all different ages – a five year old, a ten year old, a twenty year old, a thirty year old and so on, all Einsteins. The older Einsteins could tutor the young Einsteins. Think of what that clone could do for mankind (if mankind were prepared to listen). Einstein is of course dead, but more geniuses will be born and when they are and when they are recognized as such will we just let them die? Or will we clone them?

Sperm and ova banks

Less futuristic is the suggestion by Muller that we might collect sperm and ova from geniuses and establish 'banks' to be used as and when required. Sperm can be frozen and stored indefinitely and such sperm can be used as required for insemination. The process is used extensively in cattle breeding. In Denmark for example 98% of all calves are sired by AID (artificial insemination by donor). The system is also used for humans. It has been estimated that there are some 5000–10,000 AID children a year in the USA and perhaps 1000 in Europe.

So far only male sperm have been given, stored and used but in future it may be possible to remove the ovum or egg from a woman, fertilize it in a test tube and then replace the fertilized ovum in the uterus of the female. These processes have already been carried out in animals. Top quality ewes are inseminated by sperm from top quality rams. The resulting small embryos are flushed out and implanted into rabbits which act as carriers. One small rabbit can carry several embryos and the rabbits are sent, say, from Australia to America. At their destination the rabbits are killed, the embryos removed and then implanted into run-of-the-mill ewes which act as 'incubators' and eventually give birth to top class lambs.

Not many people would like to see such a system adopted for humans. There would indeed be massive public reaction to such a suggestion. Attitudes change however and modifications of this technique might become acceptable in the millennia that lie ahead.

Cultural evolution

Most scientists believe that the evolution of Man is an ever continuing process but few are willing to predict what the future 'shape' of Man is going to be. A massive increase in brain size with a corresponding decrease in the importance of the body? A world dominated by machines?

Some scientists believe that as far as Man is concerned his biological evolution has run its course – 'during the last fifty thousand years we have no evidence of an appreciable increase in mental ability'. Cultural evolution has taken over. Cultural activities require a powerful brain and in this respect no animal comes near to Man. The Man-like creatures of 30 million years ago had the brain volume equivalent to that of a new-born baby (300–400 centimetres). Within the following 29 million years the brain size almost doubled to 600 cubic centimetres as in the Australopithecines. In the following 500,000 years it more than doubled again to 1400–1700 cubic centimetres in *Homo sapiens*. The greatest growth has been in the temporal lobes of the frontal brain controlling speech and abstract thought so that Man, unlike the apes or any other animals, has a new dimension; he can think into the future. He can plan, have aims, and purposes. As a result he also has traditions and he accumulates knowledge over successive generations. He is unique in that information can be passed on, not only by genes but by words. Man has the ability, possessed to only a small degree by some other animals, to pass on his experiences and traditions by word of mouth, by books and by other technological aids. Without this added dimension, which is cultural, he might have ceased to exist because he had run his biological course and he would perhaps have been succeeded by some, at present unrecognizable, organism waiting in the shadows to take over as the primitive mammals took over and evolved from the reptiles. The main difference between the reptiles and ourselves in this respect however is that we are aware (through our culture) of such possibilities and we can be constantly on the lookout to forestall such happenings. In this way by the development of his culture, Man could go on almost for ever.

The major factor in the development of this culture has been books and printing and they have an interesting evolutionary history of their own. The first books were probably produced in China about 5500 years ago written on long wooden tablets. Paper was invented in China too by Tsai-Lun about 2000 years ago but before this the Mesopotamians used baked clay and the Egyptians used papyrus. The early Muslims scratched the Koran on skins and flat bones, India's literature was recorded on palm leaves, and Russia's on birch bark. Movable type printing was also invented by the Chinese more than 900 years ago but it was more than 400 years later before this invention was taken up by the Europeans. It is recorded that at the time that Johann Gutenberg was producing Europe's first typeset books in Germany (in the 1450s), the Koreans produced an encyclopedia of 112 volumes!

Man's future need not be determined by his environment because he can adjust it to suit himself; he does not 'need' to mutate to suit his environment. In the cold climate Man does not 'need' to develop mutants with furry skins to survive. He wears the furry skins of other animals. He does not 'need' a dark skin to survive in the tropics as he can devise suitable clothing and housing. The removal of 'natural selection' (and Man has almost achieved this) has its dangers. 'Nature' will strike back. The main danger is the almost limitless growth of population. It cannot continue unabated or natural selection will re-assert itself. Will

Man have enough sense to recognize this? His destiny could be shaped by the acts of Man.

The drifting continents

As Man struggles with his biological cultural evolution however many changes will be taking place in the physical world around him. As we saw in Chapter 3 the present shape and position of the continents is due to their drifting 'on a sea of magma'. This like all natural happenings is a continuing process. It does not stop simply because we have noted it any more than evolution stopped because Darwin drew our attention to it.

Robert Dietz and John Holden speaking to Samuel Matthews of *National Geographic* have postulated what will happen in the next 50 million years if present rates of drift continue. 'Los Angeles, far north of San Francisco by then, will nearly have reached the Aleutian Islands. Ten million years more and it will either slide into the Aleutian Trench or become part of a new landmass. Central America will be gone. The Atlantic and the Pacific will meet in the Caribbean Sea. East Africa will have broken from the rest of the African continent and the Great Rift Valley will be a long water-filled gulf. The Red Sea will have widened, the Mediterranean shrunk. Australia will have moved north beyond the latitude of Singapore and will have overrun most of Indonesia.' The process will be so gradual of course that people (if there are people) will not notice.

The life of the Sun and the solar system

The energy from the Sun is derived from the conversion of hydrogen into helium and since this goes on at a steady pace we can predict, with some degree of certainty, how long the Sun is going to last. Although it is losing about four million tons in weight every second it will go on producing energy for the next 16,000 million years which is certainly a long enough time to allow some pretty profound evolutionary changes to take place. Will the process end with a bang or a whimper? If the Sun is not different from other stars (and there is no reason to suppose that it is) then it will gradually lose its energy and it will fade. As its fuel is exhausted it will shrink to a tiny fraction of its former size becoming what is known as a 'white dwarf'. Such white dwarfs are not uncommon in the universe. It may then cool and no longer give forth heat, or light, and at its terminal point become a 'black dwarf' invisible in the sky. The Earth's temperature will long before this have dropped to −300°F. The Earth's atmosphere will have condensed to liquid air covering the eternal ice of the oceans to a depth of 30 feet. No stars will be visible. The sky will be everlastingly black as the cold planets revolve for eternity, in perpetual darkness around their black dwarf which in its heyday poured forth heat and light.

Another possibility however is that as it nears its end the white dwarf will become a nova, i.e. there will be a temporary massive increase in solar radiation so that the temperature on Earth will rise so high that the seas will boil. At the same time the Sun will increase in size until it becomes a red giant, its size becoming so great that it engulfs the nearer planets – Mercury, Venus, Earth and Mars. As this phase passes and the solar energy becomes exhausted so the Sun will shrink, becoming cold and dark, and with its planets it will spin for evermore in perpetual night.

Man, if he survives that long, will then either freeze or boil to death.

Is this the way that life will end? Probably not for as I have said elsewhere the universe is teeming with life and others will carry the torch. It may not even be the end of Earthman. With his advanced technology he may have become independent of the Sun and have developed his own nuclear reactors to replace it long before the 'black dwarf' stage. Or he may have left Earth long before to establish himself in artificial climates on other planets. There is no end to the ingenuity of Man and thus there may be no end to Man.

Further Reading

Theodosius Dobzhansky *Mankind Evolving.*
P. R. Ehrlich and Anne Ehrlich *Population, Resources, Environment*, London 1970.
Jesse L. Greenstein 'Dying Stars', *Scientific American*, January 1959.
Gerald Leach *The Biocrats*, London 1970.
Samuel W. Matthews 'This Changing Earth', *National Geographic*, January 1973.
J. D. Roslansky (Ed) *Genetics and the Future of Man*, North Holland Publishing Company, 1966.
Gordon Rattray Taylor *The Biological Time Bomb*, London 1968.
Gordon Rattray Taylor *The Doomsday Book*, London 1970.
Your Environment. Four issues per annum.

Geological Eras, Periods and their Plants and Animals

ERA	PERIOD		DURATION (millions of years)	TIME SINCE BEGINNING (millions of years)	SOME PLANT AND ANIMAL DEVELOPMENTS
CAINOZOIC (Age of Mammals)	Quaternary	Recent Pleistocene	1+	1+	Races of Man. First Man.
	Tertiary	Pliocene Miocene Oligocene Eocene/ Paleocene	69	70	Ape-like and man-like creatures. Monkeys and ancestors of apes and men. Spread of Birds. Modern mammals and flowering plants.
MESOZOIC (Age of Reptiles)	Cretaceous		65	135	Spread of flowering plants. Extinction of dinosaurs. Egg laying and marsupial mammals.
	Jurassic		45	180	First flowering plants. Dinosaurs dominant, first primitive birds, insects abundant.

	Triassic	45	225	Cone-bearing trees. Flying and water reptiles, dinosaurs, mammals like reptiles.
PALAEOZOIC	Permian	45	270	Cone-bearing trees, reptiles abundant.
	Carboniferous	80	350	Forests of tree ferns and lycopods; amphibians abundant; first reptiles.
	Devonian	50	400	Ferns and lycopods; cartilaginous and first bony fishes, lung-fishes.
	Silurian	40	440	First land plants; jawless fish.
	Ordovician	60	500	Corals; first vertebrates.
	Cambrian	100	600	Sponges, corals, marine worms, seaweeds.
ARCHEOZOIC		(approx)	(approx)	Algae and some marine worms.
PROTEROZOIC	Pre-Cambrian	4,400	5,000	Fossils rare, but some algae have been found which are approximately 2,600 million years old.

Note: This table is compiled in part from tables in Alfred S. Romer *The Procession of Life*, Weidenfeld and Nicolson.

Index